The Changing Geography of National Parks and Protected Areas

Joe Weber · Selima Sultana
Editors

The Changing Geography of National Parks and Protected Areas

palgrave
macmillan

Editors
Joe Weber
Department of Geography
University of Alabama
Tuscaloosa, USA

Selima Sultana
Department of Geography, Environment, and Sustainability
University of North Carolina Greensboro
Greensboro, NC, USA

ISBN 978-3-031-74652-9 ISBN 978-3-031-74653-6 (eBook)
https://doi.org/10.1007/978-3-031-74653-6

This Palgrave Macmillan imprint is published by the registered company Springer Nature Switzerland AG
The registered company address is: Gewerbestrasse 11, 6330 Cham, Switzerland

Dedicated to the world's Protected Areas and those who study them

Contents

List of Contributors

Matthew Balentine Department of Geoscience, One Harrison Plaza, University of North Alabama, Florence, AL, USA

Xochizeltzin Castaneda-Camacho Department of Environmental Studies, St. Olaf College, Northfield, MN, USA

John Crowell Department of Geoscience, One Harrison Plaza, University of North Alabama, Florence, AL, USA

Lary M. Dilsaver Department of Earth Sciences, University of South Alabama, Mobile, AL, USA

Charles F. Holloman III Department of Geoscience, One Harrison Plaza, University of North Alabama, Florence, AL, USA

Seth T. Kannarr Department of Geography and Sustainability, University of Tennessee, Knoxville, TN, USA

László Mari Department of Physical Geography, Eötvös Loránd University, Budapest, Hungary

Alyssa Papantonakis Department of Urban Studies and Planning, Massachusetts Institute of Technology, Cambridge, MA, USA

Michael Pretes Department of Geoscience, One Harrison Plaza, University of North Alabama, Florence, AL, USA

Selima Sultana Department of Geography, Environment, and Sustainability, University of North Carolina Greensboro, Greensboro, NC, USA

Tamás Telbisz Department of Physical Geography, Eötvös Loránd University, Budapest, Hungary

Michael A. Urban Department of Geography, University of Missouri, Columbia, MO, USA

Joe Weber Department of Geography, University of Alabama, Tuscaloosa, AL, USA

Randall K. Wilson Department of Environmental Studies, Gettysburg College, Gettysburg, PA, USA

Yonit Yogev Independent Scholar, Olympia, WA, USA

List of Figures

List of Tables

CHAPTER 1

Introduction: The Changing Geography of Protected Areas

Joe Weber and Selima Sultana

The term 'Protected Areas' [PAs] refers to a diverse set of places that are managed for their scenic, historical, ecological, or recreational significance, and are consequently protected from private profit-generated uses that prevail outside their boundaries. Spectacular national parks might be the most visible and famous of these, but there exists a wide variety of these places around the world, ranging from local parks to international ecological reserves.

The International Union for the Conservation of Nature (IUCN) provides a useful classification scheme that specifies seven types of protected areas (IUCN 2008). The IUCN recognizes that protected areas

J. Weber (✉)
Department of Geography, University of Alabama, Tuscaloosa, AL 35401, USA
e-mail: jweber2@ua.edu

S. Sultana
Department of Geography, Environment, and Sustainability, University of North Carolina Greensboro, Greensboro, NC 27402, USA
e-mail: s_sultan@uncg.edu

J. Weber and S. Sultana (eds.), *The Changing Geography of National Parks and Protected Areas*,
https://doi.org/10.1007/978-3-031-74653-6_1

may be managed by national, state, or local governments and may use a variety of terms to describe these places and the phrase 'national park' can have very different meanings in different countries. The U.S. contains 51,018 of these places in 2024, containing just over 12.9% of the land area of the country (Protected Planet 2024). While those of the National Park Service (NPS) may be the most famous, other federal agencies have sizable landholdings, none larger than the Bureau of Land Management (BLM). The U.S. Forest Service (USFS) and Fish and Wildlife Service (FWS) also have large collections of protected areas. Many of these protected areas are also under state or local management, or under private ownership.

The first of these seven categories (Category IA) is a nature reserve designed to protect natural features or biodiversity, with visitors excluded. The IUCN lists 765 such places in the U.S., including wildlife reserves, marine reserves, and small areas within national forests set aside as research natural areas. Cahuilla Mountain Research Natural Area in California and Crocodile Lake National Wildlife Refuge Marine Protected Area in Florida are examples.

Category IB corresponds to wilderness areas in the U.S. They may allow visitors and be larger than IA but should still be largely pristine. In the U.S. there are 765 wilderness areas within the National Wilderness Preservation System managed by the National Park Service, USFS, FWS, and BLM (the IUCN counts these separately from the parks, forests, or other areas within which they lie). The largest of these comprises the wilderness within Death Valley National Park. When other lands managed by state or local governments to the same standard are counted there are 1689 IB areas in the U.S.

Category II places are national parks, large areas designed to protect natural features but with fewer visitor controls and development to support tourism. In the U.S. these are managed by the National Park Service as the scenic heart of the national park system. The IUCN lists 57 places meeting this definition, though the NPS labels 63 units as 'National Parks'. Yellowstone, the Grand Canyon, Death Valley, and Everglades are examples. Category III refers to a natural monument, designed to protect a natural feature and which may be just large enough to contain that feature. National monuments, small national parks, and historical parks in the U.S. national park system fall into this category. The IUCN lists 1007 of these in the U.S., including Mount Rushmore and Muir Woods National Monument. However, most of these are state or locally managed areas, such as Maine's Lily Bay State Park, or Lost River State Park in West Virginia.

Category IV places are managed to protect a habitat or a species. These will likely be small. The U.S. equivalent is the National Wildlife Refuge, administered by FWS. There are 973 of these places in the U.S. Bon Secour National Wildlife Refuge in Alabama and Ash Meadows National Wildlife Refuge in Nevada are examples. Category V refers to protected landscapes or seascapes, reflecting a mix of scenic qualities but heavy human presence. The IUCN lists 34,063 of these in the U.S., making up more than two thirds of all protected areas in the country. This includes many units of the national park system but also state, county, and city parks. Kalaupapa National Historical Park in Hawaii, Gila Box Riparian area in Arizona (managed by the BLM), Shawnee State Forest in Ohio, and Mount Blue State Park in Maine are all examples. Finally, Category VI places are those in which natural resources are managed for extractive uses but in a sustainable fashion. National Forests would be a U.S. equivalent, though there are also state forests, waterfowl production areas, and similar units within the 481 places in the country. Rosecrans Bog in Pennsylvania, Turtle Island Wildlife Management Area in South Carolina, and Ransom County Waterfowl Production Area in North Dakota are examples.

It can safely be said that the study of protected areas has emerged as a distinct subfield. A step forward in the geographic examination of protected areas was taken when The Protected Area Specialty Group (PASG) of the Association of American Geographers (AAG) was established in 2021. This provides a community for professional geographers working on these topics. The aims of this group are to:

"promote and encourage research, teaching, and service in the field of protected areas, which encompass the creation and management of parks, forests, monuments, historic sites, wildlife refuges, conservation areas, preserves, wilderness, sacred spaces, and other landscapes, buildings, or places protected by law or custom in the United States, other countries, and international areas on Earth and other planets. This shall be accomplished through: (a) the sponsorship of paper and discussion sessions at the regional and national meetings of the American Association of Geographers, as well as other symposiums and conferences whose theme relates to protected areas; (b) the encouragement of junior scholars by hosting paper and poster competitions and awarding research grants; (c) the exchange of information among PASG members; and (d) the development of relations and interchange with other disciplines and specialty groups" (PASG 2024).

In 2023 a special issue of the *Southeastern Geographer* was devoted to the geography of protected areas, the first such special issue of its kind (Weber and Kupfer 2023). This included articles on climate change impacts on coastal heritage sites in the Southeast (Corkran et al. 2023), on tree growth within protected areas (Rochner et al. 2023), management issues at Cape Hatteras National Seashore with regard to vulnerability to storms and sea level rise in the face of increased visitation, (Flynn et al. 2023), preservation benefits in the Congaree biosphere (Hutchens, Kupfer, and Gao 2023), and racial aspects of visitation at Great Smoky Mountains National Park (Sultana et al. 2023). A special issue in the *Annals of the American Association of Geographers* will also examine national parks. There are book-length treatments of federal lands (Wilson 2016), wilderness (Vale 2005) and the national park system by geographers (Weber and Sultana 2024).

Geographers are past the point of using parks only as study sites and have gone beyond to study their management economies, cultural landscapes, travel patterns, and other features. Protected Areas Geography is now an established subfield within U.S. academic geography. We have clearly made substantial progress on developing a solid body of professional research on parks and protected areas employing geographic perspectives and techniques.

The study of protected areas is an interdisciplinary subfield that is relatively new. Within the field of geography few scholars have examined national parks or protected areas, but the topic has become a rapidly growing one. As there are no geography journals that specifically focus on this topic, this book will gather some of the most recent work, most of which has been done just in the past few years. This work does not fit within a single category, and includes land degradation, the social and political geography involved in creating new national park units, visitation by underserved segments of the population, and changing names of protected areas.

The study of protected areas is a topic with a long history but is also of growing importance and poised for significant expansion in the future. The U.S. national park system is wide open to geographers seeking new topics or places to study, and the global set of protected areas offers an even vaster array of opportunities.

The present volume is a step in this direction. It is an edited collection of recent geographic work examining different aspects of the changing geography of protected areas in the U.S. and Europe. It had its origin in conference sessions at the 2023 meeting of the American Association of Geographers in Denver, Colorado, which brought together a wide range of researchers and topics. It makes use of work using methods and data as diverse as remote sensing, nineteenth survey plats and GIS, and online visitor surveys.

Outline of Chapters

Chapter 2 is a review of work by geographers on the U.S. national park system by Joe Weber and Selima Sultana. Geographers have written about this system for over a hundred years, with a recent upsurge in interest in the twenty-first century that may relate to the 2016 centennial of the National Park Service and the attention it brought to the system. Geographers have found much to study in the system. Physical geographers have long used mountainous national parks as sites to do field work on forestry, landslides, river channel change, and similar environmental changes. Here these features are protected from many human-induced changes yet readily accessible to researchers. A range of human geographers have also looked at parks, ranging across topics such as roads, boundaries, wildlife policies, as well as the growth of the system. But it is evident that there are a number of issues that could be resolved in order for geographic research to be more effective. Compared to many other subfields Protected Area geography has strikingly few standard definitions, data sources for many topics are sparse and inconsistent, and there has been too little attention given to park units outside the 'Crown Jewels', places like Yellowstone, Yosemite, or the Grand Canyon. Geographers would do well to critically examine many aspects of park boundaries, purposes, and management, as well as connections to the world beyond the park entrance.

Protected areas may have fragile environments, which may be threatened by land uses outside their boundaries. Chapter 3 is one of the three in this volume that examines protected areas outside the U.S., in this case, Mexico. Xochizeltzin Castañeda Camacho examines three widely scattered areas in the Chihuahuan Desert of northern Mexico and the loss of wildlife habitat due to land degradation. This is particularly true in arid lands where vegetation is typically sparse, and slow-growing. She employs remotely sensed imaging techniques as well as field work to measure plant cover over time. Hers is the only chapter that discusses climate change, a topic of widespread concern with protected areas, which are often at heightened risk.

Karst terrains occur in areas of limestone when it has been eroded by water, creating caves, sinkholes, and disrupted drainage patterns. These areas are often economically depressed areas because of water scarcity and poor soils and are often avoided for farming. The result is areas with relatively intact natural vegetation as well as scenic natural features such as caves and sinkholes that draw many visitors. A number of national parks have therefore been created in Eastern Europe to preserve karst areas. Tamás Telbisz and László Mari review the significance of karst terrain in several European national parks, employing surveys to identify the benefits and disadvantages of these parks to locals. Overall, these parks have positive effects for local development, but these are insufficient to overcome many of the deeply rooted social problems of these areas.

One feature that the national parks are often known for is their wildlife. National parks have played a central role in early efforts to protect and re-store endangered species around the world. In the book's only chapter that focuses on wildlife, Randall Wilson and Alyssa Papantonakis examine the topic of wolf management, using a comparative study involving the U.S. and France. Yellowstone National Park played a pivotal part in the successful reintroduction of gray wolves in the U.S., but does this offer a model for reintroducing these animals elsewhere? And while Americans are proud of such efforts, there is also much to be learned from similar efforts in other countries. These issues are examined using a comparative study of wolves and national parks in France and the U.S.

The U.S. national park system has seen tremendous growth since Yellowstone was established in 1872. The National Park Service (NPS) was created in 1916 to manage Yellowstone and the other parks that had been added by then, and today the proliferation of new parks has resulted in a system of 429 units in 2024. There are park units in every state and territory. The state of Alabama has eight units today, along with three multi-state trails that cross the state. This number is about average, but not every place proposed for inclusion in the system is successfully added. In this chapter Lary Dilsaver and Joe Weber examined NPS records to find that 72 places were unsuccessfully proposed as parks, and the stories of who proposed them and why did their efforts fail. The NPS uses three criteria to judge a proposed unit: significance to the entire nation; suitable representation of a natural, historical, or recreational category of the system; and feasibility in terms of cost, completeness of resource inclusion, as well as taking into account potential public opposition and whether another organization or agency can protect the resources. It is evident that many proposals were of substandard or inappropriate places better preserved at the state or local level, but it is unfortunate that the NPS tried repeatedly to establish a park unit to preserve the Mobile-Tensaw River and its delta, an area of tremendous biodiversity.

Many U.S. national park system units have seen their names changed over the years, often reflecting a desire to refine a name or make it more meaningful to visitors. The U.S. national park system also contains dozens of different terms to describe the resources or purpose of a park. These too can be changed. An instance when both the name and designation change is unusual, but, happened in 2018 when Jefferson National Expansion Memorial became Gateway Arch National Park. This attracted some controversy because many thought this small urban landmark was not suitable for the 'national park' label, which to many Americans suggests a large rural natural area. Seth Kannarr and Michael A. Urban examine this name change and go beyond the question of why it took place to investigate how people feel about it. They conducted a survey to gather people's views on what labels such as 'national park' mean to them and assess their opinions about Gateway's new name. While many value the Gateway Arch, they are not convinced that it should be a national park.

It is not just names that may change, but people. The country's population has experienced a shift toward minorities, but the demographics of park visitors, and especially national park service employees have not kept pace. Yonit Yogev explores the National Park Service's experience with diversity, equity, and inclusion using interviews with participants from the Park Service, partner agencies, and Black, Indigenous, People of Color (BIPOC). Participants emphasized the need for major changes in hiring practices, deeper collaboration, and cultural change within the agency. It is evident that a lack of inclusiveness and increased diversity within both the Park Service and environmental organizations threatens the future of parks and can be considered a major environmental injustice. This is an important topic for the future of the National Park Service and parks.

Protected areas such as national park units do not exist in a vacuum but overlap with many other political boundaries, making for complex political geographies. In Chapter 9, Joe Weber investigates the changing boundaries of Canyon de Chelly National Monument in northeast Arizona. This contains a set of spectacular sandstone canyons as well as ancient Native American ruins. It became a part of the national park system in 1931, and like other park units has boundaries, a visitor center, roads, and scenic overlooks. But the monument is unique in that it is not just located within the Navajo Indian Reservation, is part of their traditional homeland and cultural landscape, people, and the land is still owned by the tribe. The monument's boundaries are also unusual and closely tied to the creation of the Navajo Indian Reservation in 1868 as well as the Public Land Survey System, which subdivided land for settlement. The monument's boundaries were revised due to limitations in this survey, and the current boundaries provide for a sort of hybrid, with a surveyed area surrounding the visitor center and unsurveyed canyons. These do not strictly correspond to frontcountry and backcountry or wilderness categories, as many Navajo live within the monument. The reservation and land survey have been altered, leaving the western end of the Monument as a relic of nineteenth-century government ambitions to settle these peoples. Drawing on sources such as General Land Office survey plats, National Park Service maps, and other cartographic sources, this chapter tells a story of the monument and its various interrelated boundaries.

Technological and social changes have brought easier travel and increased park visitation, creating the park system we are familiar with today. But changes can also potentially undermine parks. One concern is that younger generations may be losing interest in visiting parks, thanks in large part to a greater interest in video games and social media and less in outdoor activities, scenery, and experiencing history. This concern is addressed in Chapter 10 by Matthew Balentine, John Crowell, Charles F. Holloman III (all members of Generation Z), along with Michael Pretes. Their task is difficult because the NPS collects very little data about visitor demographics. The authors make use of this but also engage in visitor surveys at eight western national park units, as well as interviewing park rangers. They find that video games and social media are not a deterrent to park visitation, and that younger adults are quite interested in visiting them. This is perhaps the most ideal note on which to end this volume, and offers many opportunities to continue studying the geography of protected areas in the future.

Declarations

Conflict of interest The authors have no conflict of interest to declare that are relevant to the content of this chapter.

References

Corkran, Reilly T., Amy E. Potter, and Phillip P. Schwartz. 2023. The impacts of climate change on coastal Georgia heritage tourism sites. *Southeastern Geographer* 63: 36–53.

Flynn, Michael J., Thomas R. Allen, Meaghan E. Johnson, and David E. Hallac. 2023. Coastal science for resilience and management at the Cape Hatteras National Seashore, NC, USA. *Southeastern Geographer* 63: 54–77.

Hutchens, Lilian, John A. Kupfer, and Peng Gao. 2023. Spatial patterns of conservation benefits, priorities, and barriers in the Congaree Biosphere Region. *Southeastern Geographer* 63: 98–118.

IUCN (International Union for the Conservation of Nature). 2008. Guidelines for Applying Protected Area Management Categories. https://portals.iucn.org/library/sites/library/files/documents/PAG-021.pdf

PASG (Protected Areas Specialty Group). 2024. *Protected Areas*. https://www.aag.org/groups/protected-areas/

Rochner, Maegen L., Thomas W. Patterson, Karen J. Heeter, and Grant L. Harley. 2023. Increased growth synchrony of red spruce in response to acid deposition recovery and climate change across its southern range extent, Southeastern USA. *Southeastern Geographer* 63: 78–97.

Sultana, Selima, Josh Merced, Joe Weber, Ridwaana Allen, and Gregory Carlton. 2023. Great Smoky Mountains National Park and its missing black visitors: A preliminary analysis on the hidden architecture of landscape. *Southeastern Geographer* 63: 15–35.

Vale, Thomas R. 2005. *The American Wilderness: Reflections on nature protection in the United States*. Charlottesville: University of Virginia Press.

Weber, Joe, and John Kupfer. 2023. Guest editors' introduction: Protected areas of the American Southeast. *Southeastern Geographer* 63: 6–14.

Weber, Joe, and Selima Sultana. 2024. *The parks belong to the people: The geography of the national park system*. Athens: University of Georgia Press.

Wilson, Randall K. 2016. *America's public lands: From Yellowstone to Smokey bear and beyond*. Lanham, MD: Rowman & Littlefield Publishers.

CHAPTER 2

Geographic Research on National Parks: A Review and Prospectus

Joe Weber and Selima Sultana

Geographic Work on National Parks

National parks have naturally received tremendous attention from scholars; Graf (2013) reported a total of 4150 publications on the Florida Everglades alone between 1800 and 2010. It would seem natural that geographers would write about these places. An examination of work published since 1900 in three American journals, the *Annals of the Association of American Geographers*, *Geographical Review*, and *The Professional Geographer,* confirms this. There have been 48 papers in these journals with 'national park' or the name of a national park unit in their title. This is surely an undercount, as there could be papers written that use parks

J. Weber (✉)
Department of Geography, University of Alabama, Tuscaloosa, AL 35401, USA
e-mail: jweber2@ua.edu

S. Sultana
Department of Geography, Environment, and Sustainability, University of North Carolina Greensboro, Greensboro, NC 27402, USA
e-mail: s_sultan@uncg.edu

J. Weber and S. Sultana (eds.), *The Changing Geography of National Parks and Protected Areas*,
https://doi.org/10.1007/978-3-031-74653-6_2

or as a study site but not reflected in the title. This total also does not include Fenneman's (1916) paper on physiographic regions of the U.S., which was later used by the NPS for planning future expansion into new environments.

The earliest of these papers was a 1911 report on Mesa Verde (Atwood 1911), which had just become a national park in 1906. Other early papers reported on fieldwork or recent visits to national parks, such as that by Haas (1926) also on Mesa Verde, or Blackwelder (1933), published a few months after Death Valley became a National Monument. Work on national parks has been quite erratic over time, with no national park papers having been published in these three journals in the 1950s, but increasing numbers later, with 11 in the 1990s and 12 in the 2010s. Graf (2013) noted that publications on the Everglades reached a peak from 2000 to 2002 after a massive (but ultimately failed) restoration effort received tremendous publicity, and events such as the centennial of the National Park Service in 2016 may have prompted attention to these places or encouraged authors to refer to them by name in their titles.

Early papers, such as Haas' (1926) examination of the physical geography of the Mesa Verde plateau accompanied by the ecology of its ancient inhabitants, or Stearns' (1924) brief look at the terrain and water resources of Craters of the Moon, or Peacock's (1931) reconnaissance of Lava Beds National Monument, describe early visits to what would have been little known park units at the time. A discussion of the geology and physical geography of Rainbow Bridge National Monument (Miser et al. 1923) is another example. Some examined what we would now call human–environment interactions. Colton (1932) examined the relationship between eruptions at Sunset Crater and nearby Native American communities (now represented by the archaeological sites within Wupatki National Monument), while Ives (1947) looked at the physical geography of Zion National Park and surrounding areas, with some attention to human occupancy.

Physical geography has been a popular approach to national parks. Cooper (1937) and Field (1947) examined shrinking glaciers within Glacier Bay National Monument, while Dyson (1948) did the same for Glacier National Park. Several papers examined different physical features of Death Valley, including traces of ice age lakes (Blackwelder 1933), temperature records (Court 1949), and faulting and its effect on landforms (Beaty 1961). Churchill (1981) investigated hillside slopes within Badlands National Park. Bauer and Schmidt (1993) and Dexter and Cluer

(1999) researched sand deposits along the Colorado River within the Grand Canyon. Marcus et al. (2011) examined the wood distribution along streams in Yellowstone National Park and their relation to fires and floods. Shelton (1990) investigated water flow to Everglades National Park. It is evident that parks could accommodate field study sites for many different types of physical geography studies.

Some papers examine vegetation to understand how it has changed over the last several hundred years. Vankat (1977) examined the role of fire in the landscape of Sequoia National Park, from Native American fire use through to early NPS experiments with controlled burns. Vale (1987) does this in Yosemite National Park using repeat photography, one of many such studies in the park system; Butler and DeChano (2001) carried out one in Glacier National Park. Meyer and Youngs (2017) used this methodology in Yellowstone National Park. Sawyer and Butler (2006) examined postcards for clues as to environmental change in Glacier.

Both Taylor (1990) and Parker (1992, 1993) use Lassen Volcanic National Park as a study site for studies on meadow and forest conditions, as do Baker and Weisberg (1995) for Rocky Mountain National Park. Attention is given to park management goals at the time of allowing for natural change but attempting to stop human interference in the natural landscape. Wolf et al. (2003) look at exotic plants invading meadows in Rocky Mountain National Park. Butler and Walsh (1990), Gao and Butler (1992), and Dixon et al. (1999) examined avalanche paths in Glacier National Park. Meentemeyer and Butler (1995, 1999) examined beaver dams and their impacts in Glacier. Perroy et al. (2012) examine arroyos on an island within Channel Islands National Park and their relation to human use of the island for animal grazing. Fritschle (2008) used the field notes of surveyors for the Public Land Survey to reconstruct nineteenth-century vegetation patterns in Redwoods National Park.

Works by human geographers on parks were rare in the 1960s and 1970s. Mayer (1964) looked at the creation of Indiana Dunes National Lakeshore (now National Park) and disputes over appropriate land uses for this area. O'Brien's (1966) examination of the Yellowstone National Park road network is another exception. But from the 1990s human geographers appear in these journals with increasing frequency. Wyckoff and Dilsaver (1997) examined hundreds of promotional photographs used by the Great Northern Railway to promote Glacier National Park in the early twentieth century. Youngs (2012) likewise examined hundreds of postcards of the Grand Canyon. Hurt (2010) examines the Park Service's interpretation of the Washita Battlefield National Historic Site in Oklahoma. This site has been subject to many different interpretations by competing groups and changed significantly once the Cheyenne people became active participants in that interpretation. Some papers have been very specialized and a reflection of the diverse needs and purposes these areas serve. Doherty et al. (2013) compared two models for identifying potential emergency helicopter landing sites within Yosemite National Park. Liesch (2011) carried out research with residents around Keweenaw National Historical Park, asking them about their perceptions of the park. This revealed a wide range of responses, with social class important in distinguishing how responses vary.

Boundaries of parks were rarely examined, though Hill (1965) wrote on the Chamizal boundary dispute along the Rio Grande River that resulted in a new National Memorial in Texas. In addition to studies of failed proposals or parks that were cut from the system, parks may be 'demoted' when their designation is replaced with one seen as less impressive. Parker (2010) looked at the strange case of Oklahoma's Platt National Park, which became Chickasaw NRA in 1976. Relatively few geographers have studied the National Park Service within these journals; Smith et al. (2015) examination of the NPS campaign to address climate change both within and outside the park system is one example, as is Foresta's (1995) study on wildlife policies of the Park Service.

Looking beyond the U.S. national park system other types of protected areas have appeared in these journals. O'Brien (2012) investigated racial segregation in state parks. Walker and Solecki (1999) investigated land use changes in the New Jersey pinelands. These journals also included work on protected areas in other countries, including Peru (Barker 1980), Nepal (Byers 2005), India (Chhatre, Lakhanpal, and Prasanna 2017), Tajikistan (Cunha 2017), Kenya (Daniels and Bassett 2002), Mozambique (Lunstrum 2008, 2013), South Africa (Lunstrum 2014; Buscher 2016), Canada (Nelson and Byrne 1966), Uganda (Naughton-Treves 1997), and Brazil (Stillwell 1963).

Geographers working on the U.S. national park system have also published their work in many other journals. Among geography journals the *Journal of Historical Geography* (Young 2006; Briwa 2021; Weber 2022a, b), *Tourism Geographies* (Algeo 2013; Weber and Sultana 2013a, b), *Physical Geography* (Malanson and Butler 1994; Dixon et al. 1999; Meentemeyer and Butler 1999; Butler and DeChano 2001; Resler, Butler, and Malanson 2005; Whitesides and Butler 2011), *African*

Geographical Review (Appiah-Opoku 2004), the *Journal of Cultural Geography* (Colton and Dilsaver 2005; Parker 2010), and *GeoHumanities* (Youngs 2020) are examples. *Political Geography* features many examples of work on parks, especially outside the U.S. (Dressler 2006; Ramutsindela 2017). *Geomorphology* contains many studies set within the U.S. national park system, though relatively few are authored by geographers (Allen 1998; Wilkerson and Ginger 2003; Marston et al. 2005). Other possibilities include more specialized journals such as *Photogrammetric Engineering & Remote Sensing* (Allen and Walsh 1996), *Annals of Tourism* Research (Hanink and Stutts 2002), and *Landscape and Urban Planning* (Wade, Theobald, and Laituri 2011).

Southeastern Geographer is focused on the U.S. southeast and published an early study of park visitation focused on Great Smoky Mountains National Park (Cole and Mitchell 1969). A recent special issue included papers on management issues at Cape Hatteras National Seashore with regard to vulnerability to storms and sea level rise in the face of increased visitation, (Flynn et al. 2023), preservation benefits in the Congaree biosphere (Hutchens, Kupfer, and Gao 2023), and racial aspects of visitation at Great Smoky Mountains National Park (Sultana et al. 2023). Geographers have also published directly with the National Park Service (as with Marston and Halihan 2007).

Though they are not geography journals, the *George Wright Forum* and its successor, *Park Stewardship Forum*, are important outlets for those seeking to reach NPS staff and need to be used more by geographers. Other more specialized journals exist, such as *Yellowstone Science*, to which geographers have contributed (Steingisser and Marcus 2009). Young and Dilsaver's (2011) investigation of international projects done by the National Park Service and Weber's (2016) look at former park units are among the only geographers that have published in the *George Wright Forum*. Other journals geographers have published in include *Environmental History* (Young 2009; Youngs 2011), *Ecology and Society* (Golden Kroner et al. 2016), *Environment and History* (Young, MacEachern, and Dilsaver 2021), and *Society and Natural Resources* (Youngs et al. 2008). A special issue of the *Journal of Arizona History* was dedicated to the centennial of the Grand Canyon as a National Park, though with only one contribution by a geographer (Youngs 2019). Given the number of similar centennials approaching in the 2020s and 2030s these should present many opportunities for geographers to contribute (National Park Service 2023a, b).

Lisa Benton-Short (1998, 2016) has written excellent studies of two of the most politically complex park units: the National Mall in Washington, DC, and the Presidio in San Francisco. Dilsaver has contributed books on the management of Channel Islands (2023a), Joshua Tree (2017), Cumberland Island (2004), and Sequoia-Kings Canyon (Dilsaver and Tweed 1991), as well as a compilation of laws and documents on the national park system (1994b). Ronald Foresta has written books on the Land Between the Lakes recreation area (2013), NPS wildlife policies (1995), and the national park system and its management (1984). W. Andrew Marcus was the lead author of the monumental *Atlas of Yellowstone* (Marcus et al. 2012), a stunning showcase of maps and geographical perspectives. Weber and Sultana (2024) wrote the first thematic treatment of the human geography of the national park system, including growth and expansion, boundaries, economic geography, transportation, visitation, and a survey of park units not just within the U.S. but similar protected areas around the world, as well as prospects for other worlds. Book-length treatments of America's wilderness system (Vale 2005) and the entirety of the nation's federal lands have appeared (Wilson 2016).

It is also evident that many researchers have specialized in a particular park. Katie Algeo's work on Mammoth Cave (2004, 2005, 2010, 2013; Algeo et al. 2011) is a good example; this has examined the historical geography of the park and the effect of its establishment. Yolonda Youngs has worked extensively on the Grand Canyon (2011, 2012, 2019) and on Canyonlands (Mumich et al. 2020; Pashibin et al. 2022) as well as the visitor experience in other parks Youngs (2017, 2018). Thomas Value has written extensively on Yosemite (Vale 1987, 1994; Vale and Vale 1998). David Butler has examined the treeline of Glacier (Butler and Malanson 1994; Malanson and Butler 1994; Resler et al. 2005; Butler et al. 2009a, b; Whitesides and Butler 2011; Stine and Butler 2015), effect of wind on the landscape (Butler and DeChano 1998), avalanches (Butler and Walsh 1990; Butler et al. 2009a, b; Butler et al. 2010), other features of the physical landscape (Butler and Schipke 1992; Butler and Malanson 1999), attacks by grizzly bears (DeChano and Butler 2002), as well as fire lookouts within the park (Butler 2014). Ronald Foresta has authored numerous studies including a study of the national park system and its management (1984a), urban units (1984b), the Appalachian Trail (1987a); and the Canadian park system (1985, 1987b). W. Andrew Marcus has worked on geomorphologic impacts of mining and fires on rivers and geysers in Yellowstone (Stoughton and Marcus 2000; Marcus et al. 2001, 2002; Legleiter et al. 2003; Steingisser and Marcus 2009), and is the lead author of the *Atlas of Yellowstone* (Marcus et al. 2012). Among geographers John Kupfer has perhaps published the most varied body of work on national parks and other protected areas, with studies on tree mortality (Allen and Kupfer 2001), biotic diversity (Franklin et al. 2004), conservation benefits (Hutchens et al. 2023), ecological succession (Kipfmueller and Kupfer 2005), and the use of new technologies to map visitation and visitor activities (Kupfer et al. 2021) in national park units, state parks, and state and national forests. Selima Sultana has examined race and visitation to the national parks, especially Great Smoky Mountains (Sultana et al. 2023). Joe Weber has written on movie production in Death Valley National Park (2013, 2018a) as well as the historical geography of Lake Mead National Recreation Area (2018b, 2021, 2022a).

But in terms of published output, Lary Dilsaver has been by far the leading geographer writing on park. This includes major studies of Channel Islands (2021, 2023a), Joshua Tree (2017), Cumberland Island (2004), and Sequoia and Kings Canyon (Dilsaver and Tweed 1991). He has compiled many important documents of the National Park Service (1994a; b); carried out surveys of failed park proposals in California (2008), Montana (Dilsaver and Wyckoff 2009), and Alabama (this volume). He is also among the few that have examined political geographic aspects of parks (Dilsaver and Wyckoff 2005) and made significant contributions to understanding of park management (1987, 1990, 1994a, 2016, 2023b; Dilsaver and Strong 1990; Dilsaver and Young 2007; Colton and Dilsaver 2010).

Research Issues with National Parks

Despite the work done on national parks there are still many knowledge gaps and topics needing more research. Many of these can be seen in the context of the U.S. national park system. One obvious gap for geographers is that there has been little or no academic work on the history of National Park Service cartography. The service has a long tradition of map-making, and their maps are among the best known in the country, yet they have been ignored by scholars. The political geography of parks deserves greater attention. How can national park units in the U.S. develop greater resilience to federal shutdowns and avoid becoming pawns in ideological political battles that will likely become more common in the future? American geographers have done little with the place of indigenous peoples in parks, or the conflict between 'Yellowstone' and 'Uluru' models of park design. Congestion and crowding are topics offering tremendous opportunities for geographers, given that it often arises in geographically restricted canyon (or canyon rim) parks. Given that many congested parks have specific geographic configurations, such as canyon environments and geographers could surely offer many insights.

There are also several issues that geographers need to think about in our work on protected areas. We can think of these as particular problems of working with these types of places, at least within the U.S. Some are outdated ideas that we may bring with us when we approach these places, while others arise from the way these areas are managed. They are worth considering individually. These are phrased with reference to national parks, but similar issues apply to other types of protected areas.

The Uniqueness Problem

Yes, national parks are unique, just as every city, every county, and every census tract is unique. But they also represent a set of places operated by one agency with uniform standards for which data is available. There is far more commonality in how parks are managed than there is in how counties are administered or how schools teach geography. To move forward with park studies it is necessary to get past thinking of parks only as unique entities that cannot be compared or pooled in pursuit of a larger study. Each park may provide a data point in a statistical study. Imagine doing a study at the county level and having to explain in detail what a county is, and having to argue that even though counties have different areas and populations, and their legal powers differ among states, they can still be treated as equivalent units. This is utterly unthinkable in urban geography, yet all too common with studies of the national park system.

Most work on parks has been exclusively directed at a particular unit of the system that provides an example of a phenomenon or an ideal place to carry out research. Some work has compared several parks, as with Dilsaver and Wyckoff's (2005) investigation of park boundaries. Studies of failed park proposals (Dilsaver 2008; Dilsaver and Wyckoff 2009) or those removed from the system (Weber 2016, 2022a, b) allow for some comparative studies of the merits of different places. Weber and Sultana's (2013a, b) examination of park visitation, accessibility, and population distribution is among the few that examined a sample of parks in a comparative study.

We also need more comparative studies of parks, BLM land, national forests, wildlife refuges, and other types of protected areas. Frantz's (1998) work on Indian Reservation serves as an example of how this might be done.

The Definition Problem

Researchers must come up with a standard set of definitions about parks, facilities, visitors, and other phenomena we are interested in. While we have IUCN categories to facilitate comparison between countries (IUCN 2008), it is still possible to experience confusion, even when referring to areas managed by a single agency. Does my usage of 'park' match yours? In the U.S. 'park' may refer to a National Park, one of several dozens of types of units in the national park system, or as a generic reference to any unit of the system, and even a county or city park.

Perhaps one of the trickiest concepts is wilderness. Many large national parks in the U.S., including Yellowstone and the Grand Canyon, have no wilderness. Not one square foot or meter of it if wilderness is defined as those areas designated as such by Congress under the Wilderness Act (which is held to be the U.S. equivalent of IUCN category IB). Yet we routinely think of these parks as containing vast wildernesses where few go. This clearly requires discussion when we examine these places. Front-country and backcountry are similar terms that are perhaps more often used than understood and can be expected to mean different things to different people.

The 2019 *National Park Visitor Spending Effects* study (National Park Service 2020) identified gateway areas as everything within 300 miles (483 km) of a park unit, which includes much of the country. While this may be appropriate for what the NPS was doing, it is unlikely to fit our notion of what a gateway town is. We need more work on gateway towns and the economic geography of park units. Fletchall (2013) examined Pigeon Forge, outside Great Smoky Mountains, while Hornsby (1993) looked at Bar Harbor, Maine, but there are many other places to look at.

We are familiar with these sorts of issues in other fields. In urban geography there are a variety of ways to approach the definition of a city, including a functional approach such as a Metropolitan Statistical Area

(MSA) or one based on built-up density (the Census Bureau's urbanized areas, for example). One method unlikely to be used today is simply to define cities only by their municipal boundary; this would exclude suburbs, unincorporated areas, or 'sprawl', as well as exurbs. No serious researcher would do this now for cities, yet with parks it is routine when we adopt political boundaries without examining why they exist, what is within them, what is excluded, and what phenomena it is that we are really interested in. What might the equivalents of a metropolitan area defined by commuting or economic integration be for national park units? Why do we not already have these sorts of regions to work with?

Geographers should be prepared to approach the park system more critically. For example, why are there 429 units (or however many listed in the official total as you read this)? Where does that number come from, and what does that mean? Some of these units are not open to the public and many are adjacent and have similar resources. Others combine very different features within the same park. The NPS combines many units for management. Why are Klondike Gold Rush National Historical Park in Alaska and Seattle two different units, while the Colorado River is split into many different units? How would it affect our understanding of these places? Other units could be very well treated as entirely separate. Devils Hole is an outlier of Death Valley National Park, a tiny spring surrounded by 40 acres (16.19 hectares) of land but many kilometers from the remaining 3,000,000 acres (1,214,057 hectares) of the park. It may be more useful to discuss Devils Hole as having more in common with other small single-resource units (such as Devils Postpile or Rainbow Bridge), and perhaps it should be treated as such. Each park is treated as an inviolate whole, yet they can easily be broken down for study. Why have geographers never examined this?

We should perhaps approach park units with less concern for official park service definitions, and instead develop our own system based on our own needs. A study on a feature of several parks might omit some areas from some parks and add in relevant parts to other units. Parks need not be treated as having fixed boundaries but as a collection of resources, some of which we can draw on for a study. The North and South rims of the Grand Canyon might be treated as different units (as in Weber and Sultana 2013a, b); from a human geography perspective this makes more sense. The Forest Service does things quite differently with national forests (Thompson 2007; USFS 2012), and comparative studies would again be helpful.

The 'Crown Jewel' Problem

There is still far too much reliance on the 'crown jewel' parks, such as Yosemite or Yellowstone, as representatives for all parks, or even a lack of awareness that other units exist.

Yellowstone is certainly famous, perhaps the most famous protected area in the world, yet in many ways the least typical of all of them, and many other units face far greater challenges from climate change or external development. What happens there may not matter to the rest of system, and the rest of system may be dealing with crises that Yellowstone does not have to worry about sea level rise, coastal erosion, undocumented migrants encroaching urban populations, drought, salt cedar infestation, hurricanes, etc. It is not one of the ten biggest units or the ten most visited. The fact that merely suggesting this is itself shocking or offensive shows how deeply the crown jewel syndrome has taken a stranglehold on our understanding of the park system. We need to look beyond these units to the entire range of park units.

The Data Problem

Data remains a major bottleneck for park studies. We must find, examine, critique, and develop geospatial data on parks. One of the most basic pieces of information about a park unit is its boundaries. These can be found using several standard data sets. Parks and other protected areas in the U.S. can be mapped in Geographic Information Systems (GIS) using the United States Geological Survey (USGS) Protected Areas Database, or PAD-US, which contains the boundaries of each such place (USGS 2023). The World Database on Protected Areas (WDPA), produced by the United Nations, provides GIS data for the entire world except Antarctica, but including oceans (Protected Planet 2024). The Secretariat of the Antarctic Treaty provides digital maps (but not GIS data) for Antarctic protected areas (Secretariat of the Antarctic Treaty 2024).

While the NPS does have detailed visitation data for the system and individual parks (National Park Service 2023a, b), there are only limited demographic or geographic data in the form of visitor surveys with the exception of only a handful of parks though they are often limited by small sample size. The SAR dashboard provides a glimpse into parks, but deaths and injuries are much harder to come by (National Park Service 2018). The availability of Park GIS data is inconsistent and often quite lacking. Road networks are also difficult to obtain and may not align with those in widely used databases, such as the Census Bureau's TIGER.

Most importantly, much of the data geographers regularly need to use is lacking entirely. For example, there are no datasets on former boundaries. There is no publicly available source for law enforcement jurisdictions or activities. Land ownership data was once available online but then removed from the NPS website. There is little to no economic data beyond visitor surveys or the park spending report (National Park Service 2020). Basic questions such as the resident population of parks, the level of traffic on park roads, or the number and outcome of traffic accidents cannot be answered except for a few parks and often only for a single point in time. These are questions regularly asked of towns or cities, for which comprehensive data can be quickly accessed.

The organization of data can be frustrating. In line with the definition problem noted above, park data is often organized by parks, but inconsistently. Sometimes Sequoia-Kings Canyon is listed in a part database as one park (as it is managed within the NPS), sometimes as two. Klondike Gold Rush National Historical Park, which is managed as two parks, could appear as one or two. North Cascades may appear as one or three parks. National preserves might be treated separately from their National Park companions (as with Denali or several others with Alaska).

The Isolation Problem

National park units are land uses within the world of economic activity, political struggles, pollution, and crime. They are not magical places separate from the outside world. While in the U.S. until recently many parks were quite remote, this is no longer the case. The park system has become increasingly urbanized and integrated with the rest of the country, even serving as growth poles for population and economic growth. We need work that integrates parks with the outside world. This may be one of the most urgent needs for park research, given the vast range of external threats many parks and other protected areas face. A study of Great Smoky Mountain National Parks that ignores Gatlinburg and Pigeon Forge, or one on the west unit of Saguaro that ignores all the homes just outside the park and the commuters zipping to work along park roads, would be seriously inaccurate. There is a tremendous amount of commuting going on in and through national parks. Cell towers and cell coverage in parks will surely increase, threatening park values. There is also a growing effort to make parks safer; increasing cell coverage may be necessary for this, as are redesigned roads and more guardrails.

Urban parks have problems that large rural areas do not have. Harry S Truman National Historic Site includes a detached unit preserving his family farmhouse. This tiny unit is now in a suburban area of Kansas City and is bounded by a Burlington Coat Factory store, Sonic Drive-In, and Grandview Chinese Buffet restaurants, along with apartments and an office park. This is quite different from a western park bounded by national forests, Indian reservations, and wilderness, but is perhaps even more of a management challenge for the NPS. While visiting the farmhouse visitors will not just see these neighbors intruding on the rustic farm setting, they will hear and smell them! Geographers are accustomed to working with viewsheds to identify visible areas, and perhaps this can be expanded to include other senses. In 2023 the media gave attention to retail chains or restaurants that were closing certain locations due to high crime that disrupted their business and chased customers away; in this way these companies can use geography to avoid problems and maintain their customer base. The National Park Service does not have this option.

The Representativeness Problem

A park may or may not contain a representative of a particular style of architecture, ecosystem, battlefield, or other feature, but, more importantly, we as researchers are definitely not representative of the average park visitor. We are more educated, often higher income, and surely far more aware of the history and nature a particular park was created to preserve. Yet all too often we forget this and assume that all park visitors share our perspective and must value parks as we do. We are then naturally shocked and appalled when we hear that others may have entirely different valuation for national park lands or would prefer to use them that we feel is entirely incompatible. This in turn can lead to a breakdown in communication, increased polarization, and failure.

This is especially likely given that park supporters are likely to live far away from parks, and quite possibly in a large urban area. Our way of life is quite different from those who live in and around parks. If you don't ride a dirt bike, own cattle, or prospect for gold the creation of a new park unit will probably have no effect on your behavior. Those who do will see a substantial reduction in their hobbies or economic livelihood. City people might consider their own reaction if their favorite urban green space were to ban bicycles, dogs, and wireless devices, or required a federal permit to send and receive work-related emails within the park, with violators subject to prosecution in federal court. An outrageous overreach of federal authority and an affront to the rights of people to live and work? Yes, but only for some; those living in rural areas near parks will not see any effect on their behavior. Parks are a bridge between an increasingly divided America, but only if those on both ends are willing to meet in the middle. Recognizing that we are not representative, and that there are different points of view, is a start.

The 'Everything Has Already Been Said' Problem

Perhaps the greatest problem is a belief that everything that can be said about a park has already been said or written. This is easy to encounter when browsing online or in a bookstore and seeing the volume of writing on a park. Yet anyone who has dug into a particular topic or even a single park will probably be amazed how little of scholarly interest has been written about it. Most park units have rarely or never been written about by academic scholars and certainly not by geographers. Allied disciplines are no different; there have been no comprehensive archaeological surveys at many parks. Neither do most universities offer classes on national parks.

There are many topics about which we know very little. The interaction between the NPS and urban change needs to be examined. Many units are in urban areas, but with very different circumstances. William Howard Taft National Historic Site preserves the former president's home on the edge of downtown Cincinnati. There was once a neighborhood of large homes that has since declined, with a hospital and other new buildings taking over. One consequence was that the park service had land to build a visitor center and parking. Independence National Historical Park was part of an urban renewal project in downtown Philadelphia. At Harry S Truman National Historical Site in suburban Kansas City the NPS carefully preserved his home within an intact and stable residential neighborhood, minimizing impacts on the community. Will new units such as Medgar and Myrlie Evers Home National Monument, located in a suburban home at the end of a dead-end street, have a positive or negative effect on their surrounding neighborhoods? Is the park service an agent for neighborhood stability or decline? The impact of park units on urban areas needs to be examined.

The Future of Geographic Research on National Parks

Although geographers have been examining and writing on national parks for over a century the amount of scholarship has only recently begun to dramatically increase, and anything like a critical mass of scholars has come into being. It is to be hoped that geographers will take up the many opportunities to continue to do great work on parks, including contributing to an understanding of visitor patterns, transport solutions, regional economic impacts, mapping, and addressing the myriad environmental conundrums of managing heavily visited places.

Declarations

Conflict of interest The authors have no conflict of interest to declare that are relevant to the content of this chapter.

References

Algeo, Katie. 2004. Mammoth Cave and the making of place. *Southeastern Geographer* 44: 27–47.

Algeo, Katie. 2005. Mammoth Cave National Park and rural economic development. In *Rural change and sustainability: Agriculture, the environment and communities*, ed. S.J. Essex, et al., 230–243. Wallingford, Oxfordshire: CABI Publishing.

Algeo, Katie 2010. Remembering rural community: historical GIS for Mammoth Cave National Park. In *Geographical Perspectives on Sustainable Rural Change*, ed. Dick Winchell, et al., 403–413. Brandon, Manitoba: Rural Development Institute.

Algeo, Katie. 2013. Underground tourists/tourists underground: African American tourism to Mammoth Cave. *Tourism Geographies* 15: 380–404.

Algeo, Katie, Ann Epperson, and Matt Brunt. 2011. Historical GIS as a platform for public memory at Mammoth Cave National Park. *International Journal of Applied Geospatial Research* 2: 19–36.

Allen, Thomas R. 1998. Topographic context of glaciers and perennial snowfields, Glacier National Park, Montana. *Geomorphology* 21: 207–216.

Allen, T.R., and J.A. Kupfer. 2001. Spectral response and spatial pattern of Fraser fir mortality and regeneration, Great Smoky Mountains. *Plant Ecology* 156: 59–74.

Allen, Thomas R., and S.J. Walsh. 1996. Spatial and compositional pattern of alpine treeline, Glacier National Park, Montana. *Photogrammetric Engineering and Remote Sensing* 62 (11): 1261–1268.
Andrew, Marcus W., G.A. Meyer, and D.R. Nimmo. 2001. Geomorphic control on long-term persistence of mining impacts, Soda Butte Creek. *Yellowstone National Park. Geology* 29 (4): 355–358.
Appiah-Opoku, Seth. 2004. Rethinking ecotourism: The case of Kakum National Park in Ghana. *African Geographical Review* 23: 49–63.
Atwood, Wallace W. 1911. A geographic study of the Mesa Verde. *Annals of the Association of American Geographers* 1: 95–100.
Baker, William L., and Peter J. Weisberg. 1995. Landscape analysis of the forest-tundra ecotone in Rocky Mountain National Park, Colorado. *Professional Geographer* 47: 361–375.
Barker, Mary L. 1980. National parks, conservation, and agrarian reform in Peru. *Geographical Review* 70: 1–18.
Bauer, Bernard O., and John C. Schmidt. 1993. Waves and sandbar erosion in the Grand Canyon: Applying coastal theory to a fluvial system. *Annals of the Association of American Geographers* 83: 475–497.
Beaty, Chester B. 1961. Topographic effects of faulting: Death Valley, California. *Annals of the Association of American Geographers* 51: 234–240.
Benton, Lisa. 1998. *The presidio: From army post to national park*. Boston: Northeastern University Press.
Benton-Short, Lisa. 2016. *The national mall: No ordinary space*. Toronto: University of Toronto Press.
Blackwelder, Eliot. 1933. Lake manly: An extinct lake of death valley. *Geographical Review* 23: 464–471.
Briwa, Robert. 2021. The changing terrain of heritage at a Montana state park. *Journal of Historical Geography* 71: 17–27.
Büscher, Bram. 2016. Reassessing fortress conservation? New media and the politics of distinction in Kruger National Park. *Annals of the American Association of Geographers* 106: 114–129.
Butler, David R. 2014. *Fire lookouts of glacier national park*. Charleston, SC: Arcadia Publishing.
Butler, David R., and Lisa M. DeChano. 1998. Hazardous wind events in Glacier National Park, Montana. *The Southwestern Geographer* 2: 57–71.
Butler, David R., and Lisa M. DeChano. 2001. Environmental change in Glacier National Park, Montana: An assessment through repeat photography from fire lookouts. *Physical Geography* 22: 291–304.
Butler, David R., Jacob A. Maas, and Carol F. Sawyer. 2009a. Traumatic resin ducts in tree rings: Advancing tree-ring dating of snow avalanches in Glacier National Park, Montana. *Papers and Proceedings of the Applied Geography Conference* 32: 181–188.

Butler, David R., and George P. Malanson. 1999. Site locations and characteristics of miniature patterned ground, eastern Glacier National Park, Montana, U.S.A. *Landform Analysis* 2: 45–49.

Butler, David R., George P. Malanson, and David M. Cairns. 1994. Stability of alpine treeline in Glacier National Park, Montana, U.S.A. *Phytocoenologia* 22: 485–500.

Butler, David R., George P. Malanson, Lynn M. Resler, Stephen J. Walsh, Forrest D. Wilkerson, Ginger L. Schmid, and Carol F. Sawyer. 2009b. Geomorphic patterns and processes at Alpine treeline. In *The changing treeline of glacier National Park, Montana, U.S.A*, ed. D.R. Butler, G.P. Malanson, S.J. Walsh, and D.B. Fagre, 63–84. Amsterdam: Elsevier.

Butler, David R., Carol F. Sawyer, and Jacob A. Maas. 2010. Tree-ring dating of snow avalanches in Glacier National Park, Montana, USA. In *Tree rings and natural hazards*, ed. Markus Stoffel, Michelle Bollschweiler, David R. Butler, and Brian H. Luckman, 33–44. Dordrecht: Springer.

Butler, David R., and Katherine A. Schipke. 1992. The strange case of the appearing (and disappearing) lakes: The use of sequential topographic maps of Glacier National Park, Montana. *Surveying and Land Information Systems* 52: 150–154.

Butler, David R., and Stephen J. Walsh. 1990. Lithologic, structural, and topographic influences on snow-avalanche path location, eastern Glacier National Park, Montana. *Annals of the Association of American Geographers* 80: 362–378.

Byers, Alton. 2005. Contemporary human impacts on Alpine ecosystems in the Sagarmatha (Mt. Everest) national park, Khumbu, Nepal. *Annals of the Association of American Geographers* 95: 112–140.

Chhatre, Ashwini, Shikha Lakhanpal, and Satya Prasanna. 2017. Heritage as weapon: Contested geographies of conservation and culture in the Great Himalayan National Park Conservation Area, India. *Annals of the American Association of Geographers* 107: 456–464.

Churchill, Robert R. 1981. Aspect-related differences in Badlands slope morphology. *Annals of the Association of American Geographers* 71: 374–388.

Cole, Robert L.S.., Jr., and Lisle Series Mitchell. 1969. Attendance as a negative function of distance, Great Smoky Mountains National Park campgrounds. *Southeastern Geographer* 9: 13–24.

Colton, Craig E., and Lary M. Dilsaver. 2005. The hidden landscape of Yosemite National Park. *Journal of Cultural Geography* 22: 27–50.

Colton, Craig E., and Lary M. Dilsaver. 2010. The devil in the cathedral: Sewage and garbage in Yosemite National Park. In *Cities and Nature in the American West*, ed. Char Miller, 154–170. Reno: University of Nevada.

Colton, Harold S. 1932. Sunset Crater: The effect of a volcanic eruption on an ancient pueblo. *Geographical Review* 22: 582–590.

Cooper, William S. 1937. The problem of Glacier Bay, Alaska: A study of glacier variations. *Geographical Review* 27: 37–62.

Court, Arnold. 1949. How hot is Death Valley? *Geographical Review* 39: 214–220.

Cunha, Stephen F. 2017. Perestroika to parkland: The evolution of land protection in the Pamir Mountains of Tajikistan. *Annals of the American Association of Geographers* 107: 465–479.

Daniels, Robert, and Thomas J. Bassett. 2002. The Spaces of conservation and development around Lake Nakuru National Park, Kenya. *Professional Geographer* 54: 481–490.

DeChano, Lisa M., and David R. Butler. 2002. An analysis of attacks by grizzly bears (Ursus arctos horribilis) in Glacier National Park, Montana. *The Geographical Bulletin* 44: 30–41.

Dexter, Leland R., and Brian L. Cluer. 1999. Cyclic erosional instability of sandbars along the Colorado River, Grand Canyon, Arizona. *Annals of the Association of American Geographers* 89: 238–266.

Dilsaver, Lary M. 1987. The evolution of land use in Giant Forest, Sequoia National Park. *Yearbook of the Association of Pacific Coast Geographers* 49: 35–50.

Dilsaver, Lary M. 1990. Conservation conflict and the founding of Kings Canyon National Park. *California History* 69: 196–205.

Dilsaver, Lary M. 1994a. *America's national park system: The critical documents*. Lanham, MD: Rowman & Littlefield.

Dilsaver, Lary M. 1994b. Preservation choices at Muir Woods. *Geographical Review* 84: 290–305.

Dilsaver, Lary M. 2004. *Cumberland island national seashore: A history of conservation conflict*. Charlottesville: University of Virginia Press.

Dilsaver, Lary M. 2008. Not of national significance: Failed national park proposals in California. *California History* 85: 4–23.

Dilsaver, Lary M. 2016. A national park in the wasteland: American and National Park Service perceptions of the desert. *The Public Historian* 38: 38–55.

Dilsaver, Lary M. 2017. *Preserving the desert: A history of Joshua tree national park*. Staunton, VA: George F. Thompson Publishing.

Dilsaver, Lary M. 2023a. *Restoring nature: The evolution of channel islands national park*. Lincoln: University of Nebraska Press.

Dilsaver, Lary M. 2023b. Obstacles to removing non-native species from a national park. *Parks Stewardship Forum* 39: 83–90.

Dilsaver, Lary M., and Timothy Babalis. 2021. *The oceanic park: An administrative history of channel islands national park*. Ventura: Channel Islands National Park.

Dilsaver, Lary M., and Douglas H. Strong. 1990. Sequoia and Kings Canyon National Parks: One hundred years of preservation and resource management. *California History* 69: 98–117.

Dilsaver, Lary M., and William C. Tweed. 1991. *Challenge of the big trees: A resource history of sequoia and kings canyon national parks*. Tucson: University of Arizona Press.

Dilsaver, Lary M., and William Wyckoff. 2005. The political geography of national parks. *Pacific Historical Review* 74: 237–266.

Dilsaver, Lary M., and William Wyckoff. 2009. Failed national parks in the last best place. *Montana: the Magazine of Western History* 59: 91–93.

Dilsaver, Lary M., and Terence Young. 2007. U.S. parks and protected areas: Origins, meanings and management. *Historical Geography* 35: 7–11.

Dixon, Richard W., David R. Butler, Lisa M. DeChano, and Julie A. Henry. 1999. Avalanche hazard in Glacier National Park: An El Niño connection? *Physical Geography* 20: 461–467.

Doherty, Paul, Qinghua Guo, and Otto Alvarez. 2013. Expert versus machine: A comparison of two suitability models for emergency helicopter landing areas in Yosemite National Park. *Professional Geographer* 65: 466–481.

Dressler, Wolfram H., Christian A. Kull, and Thomas C. Meredith. 2006. The politics of decentralizing national parks management in the Philippines. *Political Geography* 25: 789–816.

Dyson, James L. 1948. Shrinkage of Sperry and Grinnell glaciers, glacier national park, Montana. *Geographical Review* 38: 95–103.

Fenneman, Nevin M. 1916. Physiographic divisions of the United States. *Annals of the Association of American Geographers* 6: 19–98.

Field, William O., Jr. 1947. Glacier recession in Muir Inlet, Glacier Bay, Alaska. *Geographical Review* 37: 369–399.

Fletchall, Ann. 2013. Making sense of the strip: The postmodern pastiche of Pigeon Forge, Tennessee. *Southeastern Geographer* 53: 102–122.

Flynn, Michael J., Thomas R. Allen, Meaghan E. Johnson, and David E. Hallac. 2023. Coastal science for resilience and management at the Cape Hatteras National Seashore, NC, USA. *Southeastern Geographer* 63: 54–77.

Foresta, Ronald A. 1984a. *America's national parks and their keepers*. Washington, DC: Resources for the Future.

Foresta, Ronald A. 1984b. America's urban national parks. *Urbanism past and Present* 9: 1–14.

Foresta, Ronald A. 1985. Natural regions for national parks: The Canadian experience. *Applied Geography* 5: 179–194.

Foresta, Ronald A. 1987a. Transformation of the appalachian trail. *Geographical Review* 77: 76–85.

Foresta, Ronald A. 1987b. New national parks: Lessons from the United States and Canada. *Forum for Applied Research and Public Policy* 2: 95–108.

Foresta, Ronald A. 1995. *Wildlife policies in the U.S. national parks*. Washington, DC: Island Press.

Foresta, Ronald A. 2013. *The land between the lakes: A geography of the forgotten future*. Knoxville: University of Tennessee Press.

Franklin, S.B., J.A. Kupfer, J.W. Grubaugh, and M.L. Kennedy. 2004. A multi-taxa analysis of biotic diversity in Natchez Trace State Forest, western Tennessee. *Environmental Monitoring and Assessment* 93: 31–54.

Frantz, Klaus. 1998. *Indian reservations in the United States: Territory, sovereignty, and socioeconomic change*. Chicago: University of Chicago Press.

Fritschle, Joy A. 2008. Reconstructing historic ecotones using the public land surveys: The lost prairies of Redwood National Park. *Annals of the Association of American Geographers* 98: 24–39.

Gao, Jay K., and David R. Butler. 1992. Terrain influences on total length of snow-avalanche paths in southern Glacier National Park, Montana. *The Geographical Bulletin* 34: 91–101.

Golden Kroner, R.E., R. Krithivasan, and M.B. Mascia. 2016. Effects of protected area downsizing on habitat fragmentation in Yosemite National Park (USA), 1864–2014. *Ecology and Society* 21: 22.

Graf, William L. 2013. Water resource science, policy, and politics for the Florida Everglades. *Annals of the Association of American Geographers* 103: 353–362.

Haas, William H. 1926. The cliff-dweller and his habitat. *Annals of the Association of American Geographers* 16: 167–215.

Hanink, Dean M., and Matthew Stutts. 2002. Spatial demand for national battlefield parks. *Annals of Tourism Research* 29: 707–179.

Hill, James E., Jr. 1965. El Chamizal: A century-old boundary dispute. *Geographical Review* 55: 510–522.

Hornsby, Stephen J. 1993. The gilded age and the making of Bar Harbor. *Geographical Review* 83: 455–468.

Hurt, Douglas A. 2010. Reinterpreting the Washita Battlefield National Historic Site. *Geographical Review* 100: 375–393.

Hutchens, Lilian, John A. Kupfer, and Peng Gao. 2023. Spatial patterns of conservation benefits, priorities, and barriers in the Congaree Biosphere Region. *Southeastern Geographer* 63: 98–118.

IUCN (International Union for the Conservation of Nature). 2008. *Guidelines for applying protected area management categories*. https://portals.iucn.org/library/sites/library/files/documents/PAG-021.pdf

Ives, Ronald L. 1947. Reconnaissance of the Zion hinterland. *Geographical Review* 37: 618–638.

Kipfmueller, K.F., and J.A. Kupfer. 2005. Complexity of successional pathways in subalpine forests of the Selway-Bitterroot Wilderness Area, USA. *Annals of the Association of American Geographers* 95: 495–510.

Kupfer, John A., Zhenlong Li, Huan Ning, and Xiao Huang. 2021. Using mobile device data to track the effects of the COVID-19 pandemic on spatiotemporal patterns of national park visitation. *Sustainability* 13: 9366.

Legleiter, C.J., R.L. Lawrence, M.A. Fonstad, W.A. Marcus, and R.J. Aspinall. 2003. Fluvial response a decade after wildfire in the northern Yellowstone ecosystem: A spatially explicit analysis. *Geomorphology* 54 (3–4): 119–13.

Liesch, Matthew. 2011. Partnerships and photographs: Community conceptions of Keweenaw National Historical Park. *Geographical Review* 101: 497–517.

Lunstrum, Elizabeth. 2008. Mozambique, neoliberal land reform, and the Limpopo National Park. *Geographical Review* 98: 339–355.

Lunstrum, Elizabeth. 2013. Articulated sovereignty: Extending Mozambican state power through the Great Limpopo Transfrontier Park. *Political Geography* 36: 1–11.

Lunstrum, Elizabeth. 2014. Green militarization: Anti-poaching efforts and the spatial contours of Kruger National Park. *Annals of the Association of American Geographers* 104: 816–832.

Malanson, George P., and David R. Butler. 1994. Tree-tundra competitive hierarchies, soil fertility gradients, and the elevation of treeline in Glacier National Park, Montana. *Physical Geography* 15: 166–180.

Marcus, W., James Rasmussen Andrew, and Mark A. Fonstad. 2011. Response of the fluvial wood system to fire and floods in northern Yellowstone. *Annals of the Association of American Geographers* 101: 21–44.

Marcus, W., James L. Andrew, Ann Rodman Meacham, and Alethea Steingisser, eds. 2012. *The atlas of Yellowstone*. Berkeley: University of California Press.

Marcus, W.A., R.A. Marston, C.C.R. Marcus Jr, and R.D. Gray. 2002. Mapping the spatial and temporal distributions of large woody debris in rivers of the Greater Yellowstone Ecosystem, U.S.A. *Geomorphology* 44 (3–4): 323–335.

Marston, Richard A., John D. Mills, David R. Wrazien, Beau Bassett, and Dale K. Splinter. 2005. Effects of Jackson Lake Dam on the Snake River and its floodplain, Grand Teton National Park, Wyoming, USA. *Geomorphology* 71: 79–98.

Marston, Richard A., and Todd Halihan. 2007. *Geomorphic Adjustment of the Washita River, Washita Battlefield National Historic Site, Oklahoma*. Natural Resource Technical Report NPS/NRPC/WRD/NRTR—2007/070.

Mayer, Harold M. 1964. Politics and land use: The Indiana shoreline of Lake Michigan. *Annals of the Association of American Geographers* 54: 508–523.

Meentemeyer, Ross K., and David R. Butler. 1995. Temporal and spatial changes in beaver pond locations, eastern Glacier National Park, Montana, USA. *The Geographical Bulletin* 37: 97–104.

Meentemeyer, Ross K., and David R. Butler. 1999. Hydrogeomorphic effects of beaver dams in Glacier National Park, Montana. *Physical Geography* 20: 436–446.

Meyer, J., and Y. Youngs. 2017. Historical landscape change in Yellowstone National Park: Demonstrating the value of intensive field observation and repeat photography. *Geographical Review* 108: 387–409.

Miser, H.D., K.W. Trimble, and Sidney Paige. 1923. The Rainbow Bridge, Utah. *Geographical Review* 13: 518–531.

Mumich, N., G. Buckley, Y. Youngs, and T. Martello. 2020. Protecting Canyonlands: Anatomy of a national park controversy in Southern Utah. *Social Sciences* 9: 1–21.

National Park Service. 2018. *2017 Annual SAR Dashboard*. https://nps.maps.arcgis.com/apps/opsdashboard/index.html#/b526c87ae21f4a669eb6c9238c2c4bcf

National Park Service. 2020. 2019 National Park Visitor Spending Effects: Economic Contributions to Local Communities, States, and the Nation. https://www.nps.gov/subjects/socialscience/vse.htm

National Park Service. 2023a. *STATS—Welcome to Visitor Use Statistics*. https://irma.nps.gov/Stats/

National Park Service. 2023b. *Park Anniversaries*. https://www.nps.gov/subjects/npscelebrates/park-anniversaries.htm

Naughton-Treves, Lisa. 1997. Farming the forest edge: Vulnerable places and people around Kibale National Park, Uganda. *Geographical Review* 87: 27–46.

Nelson, J.G., and A.R. Byrne. 1966. Man as an instrument of landscape change: Fires, floods, and national parks in the Bow Valley, Alberta. *Geographical Review* 56: 226–238.

O'Brien, Bob R. 1966. The future road system of Yellowstone National Park. *Annals of the Association of American Geographers* 56: 385–407.

O'Brien, William E. 2012. State parks and Jim Crow in the decade before Brown v. Board of Education. *Geographical Review* 102: 166–179.

Parker, Albert J. 1992. Spatial variation in diameter structures of forests in Lassen Volcanic National Park, California. *Professional Geographer* 44: 147–160.

Parker, Albert J. 1993. Structural variation and dynamics of lodgepole pine forests in Lassen Volcanic National Park, California. *Annals of the Association of American Geographers* 83: 613–629.

Parker, Albert J. 2010. A park of the people: The demotion of Platt National Park, Oklahoma. *Journal of Cultural Geography* 27: 151–175.

Pashibin, T., G. Buckley, and Y. Youngs. 2022. Canyonlands national park: A multiple-use test case. *Yearbook of the Association of Pacific Coast Geographers* 84: 37–65.

Peacock, Martin A. 1931. The Modoc Lava Field, Northern California. *Geographical Review* 21: 259–275.

Perroy, Ryan L., Bodo Bookhagen, Oliver Chadwick, and Jeffrey T. Howarth. 2012. Holocene and anthropocene landscape change: Arroyo formation on Santa Cruz Island, California. *Annals of the Association of American Geographers* 102: 1229–1250.

Protected Planet. 2024. *Explore the World's Protected Areas*. https://www.protectedplanet.net/en/thematic-areas/wdpa?tab=WDPA

Ramutsindela, Maano. 2017. Greening Africa's borderlands: The symbiotic politics of land and borders in peace parks. *Political Geography* 56: 106–113.

Resler, Lynn M., David R. Butler, and George P. Malanson. 2005. Topographic shelter and conifer establishment and mortality in an alpine environment, Glacier National Park, Montana. *Physical Geography* 26: 112–125.

Sawyer, Carol F., and David R. Butler. 2006. The use of historical picture post-cards as photographic sources for examining environmental change: Promises and problems. *Geocarto International* 21: 73–80.

Secretariat of the Antarctic Treaty. 2024. *Area Protection and Management*. https://www.ats.aq/e/protected.html

Shelton, M.L. 1990. Surface-water flow to Everglades National Park. *Geographical Review* 80: 355–369.

Smith, Langdon, Laura Karosic, and Elizabeth Smith. 2015. Greening U.S. National parks: Expanding traditional roles to address climate change. *Professional Geographer* 67: 438–446.

Stearns, Harold T. 1924. Craters of the Moon National Monument. *Geographical Review* 14: 362–372.

Steingisser, A., and W Andrew Marcus. 2009. Human Impacts on geyser basins. *Yellowstone Science* 17 (1): 1–18.

Stillwell, H. Daniel. 1963. National parks of Brazil: A study in recreational geography. *Annals of the Association of American Geographers* 53: 391–406.

Stine, Melanie B., and David R. Butler. 2015. Effects of fire on geomorphic factors and seedling site conditions within the alpine treeline ecotone, Glacier National Park, MT. *CATENA* 132: 37–44.

Stoughton, J., and W.A. Marcus. 2000. Persistent impacts of trace metals from mining on floodplain grass communities along Soda Butte Creek, Yellowstone National Park. *Environmental Management*. 25 (3): 305–320.

Sultana, Selima, Josh Merced, Joe Weber, Ridwaana Allen, and Gregory Carlton. 2023. Great Smoky Mountains National Park and its missing black visitors: A preliminary analysis on the hidden architecture of landscape. *Southeastern Geographer* 63: 15–35.

Taylor, Alan H. 1990. Tree invasion in meadows of Lassen Volcanic National Park, California. *Professional Geographer* 42: 457–470.

Thompson, Tom L. 2007. *A Historical Analysis of Consolidation at the National Forest and District Level in the U.S. Forest Service*. https://forestservicemuseum.org/wp-content/uploads/2020/07/Historical-Analysis-of-Consolidation-2007.pdf

United States Forest Service. 2012. *Establishment and Modification of National Forest Boundaries and National Grasslands*. https://www.fs.fed.us/land/staff/Documents/Establishment%20and%20Modifications%20of%20National%20Forest%20Boundaries%20and%20National%20Grasslands%201891%20to%202012.pdf
USGS. 2023. *PAD-US Data Overview*. https://www.usgs.gov/programs/gap-analysis-project/science/pad-us-data-overview
Vale, Thomas and Geraldine Vail. 1998. *Walking With Muir Across Yosemite*. University of Wisconsin Press.
Vale, Thomas R. 1987. Vegetation change and park purpose in the high elevations of Yosemite National Park, California. *Annals of the Association of American Geographers* 77: 1–18.
Vale, Thomas R. 1994. *Time and Tuolumne landscape: Continuity and change in the yosemite high country*. Salt Lake City: University of Utah Press.
Vale, Thomas R. 2005. *The American wilderness: Reflections on nature protection in the United States*. Charlottesville: University of Virginia Press.
Vankat, John L. 1977. Fire and man in Sequoia National Park. *Annals of the Association of American Geographers* 67: 17–27.
Wade, Alisa A., David M. Theobald, and Melinda J. Laituri. 2011. A multi-scale assessment of local and contextual threats to existing and potential U.S. protected areas. *Landscape and Urban Planning* 101: 215–227.
Walker, Robert T., and William D. Solecki. 1999. Managing land use and land-cover change: The New Jersey Pinelands Biosphere Reserve. *Annals of the Association of American Geographers* 89: 220–237.
Weber, Joe. 2013. Star Wars in Death Valley. In *Proceedings of the 9th Death Valley Conference on History and Prehistory* 9: 8–23.
Weber, Joe. 2016. America's lost national park units: A closer look. *The George Wright Forum* 33: 59–69.
Weber, Joe. 2018a. On Location: The geography of film and television in Death Valley. In *Proceedings of the 10th Death Valley Conference on History and Prehistory* 10: 178–195.
Weber, Joe. 2018b. The Colorado and virgin rivers before lake mead. *Journal of Maps* 14: 583–588.
Weber, Joe. 2022a. *Mapping historical Las Vegas: A cartographic journey*. Reno: University of Nevada Press.
Weber, Joe. 2022b. America's lost national monuments. *Journal of Historical Geography* 78: 55–68.
Weber, Joe, and Matthew LaFevor. 2021. Exploring remnant landscapes of Nevada's Arrowhead Trail. *Geographical Review* 111: 437–457.
Weber, Joe, and Selima Sultana. 2013a. The Civil Rights Movement and the future of the national park system. *Tourism Geographies* 15: 444–469.

Weber, Joe, and Selima Sultana. 2013b. Why do so few minority people visit national parks? Visitation and the accessibility of 'America's Best Idea.' *Annals of the Association of American Geographers* 103: 437–464.

Weber, Joe, and Selima Sultana. 2024. *The parks belong to the people: The geography of the national park system*. Athens: University of Georgia Press.

Whitesides, Clayton J., and David R. Butler. 2011. Adequacies and deficiencies of alpine and subalpine treeline studies in the national parks of the western USA. *Progress in Physical Geography* 35: 19–42.

Wilkerson, Forrest D., and Ginger L. Schmid. 2003. Debris flows in Glacier National Park: Geomorphology and hazards. *Geomorphology* 55: 317–328.

Wilson, Randall K. 2016. *America's Public Lands: From Yellowstone to Smokey Bear and Beyond*. Lanham, MD: Rowman & Littlefield Publishers.

Wolf, Joy J., Susan Beatty, and Greg Carey. 2003. Invasion by sweet clover (melilotus) in montane grasslands, Rocky Mountain National Park. *Annals of the Association of American Geographers* 93: 531–543.

Wyckoff, William, and Lary M. Dilsaver. 1997. Promotional imagery of Glacier National Park. *Geographical Review* 87: 1–26.

Young, Terence. 2006. False, cheap and degraded: When history, economy and environment collided at Cades Cove, Great Smoky Mountains National Park. *Journal of Historical Geography* 32: 169–189.

Young, Terence. 2009. A Contradiction in Democratic Government: W. J. Trent, Jr., and the struggle to desegregate national park campgrounds. *Environmental History* 14: 651–682.

Young, Terence, and Lary M. Dilsaver. 2011. Collecting and diffusing "the world's best thought": International cooperation by the National Park Service. *George Wright Forum* 28: 269–278.

Young, Terence, Alan MacEachern, and Lary Dilsaver. 2021. From conservation to cooperation: A history of Canada-US national park relations. *Environment and History* 27: 607–634.

Youngs, Yolonda. 2011. On Grand Canyon postcards. *Environmental History.* 16: 138–147.

Youngs, Yolonda. 2012. Editing nature in Grand Canyon National Park postcards. *Geographical Review* 102: 486–509.

Youngs, Yolonda. 2017. Constructing place attachment in Grand Teton National Park. In *Explorations in place attachment: A geographic perspective*, ed. Jeffrey S. Smith, 117–132. New York: Routledge.

Youngs, Yolonda. 2018. Wild, unpredictable, and dangerous: a historical geography of hazards and risk in U.S. national parks. In *The American Environment Revisited: Environmental Historical Geographies of the U.S.*, Geoffrey Buckley and Yolonda Youngs, 59–80. Lanham: Rowman and Littlefield Publishers.

Youngs, Yolonda. 2019. Viewing power and place at the Grand Canyon. *Journal of Arizona History* 60: 507–524.
Youngs, Yolonda. 2020. Danger beyond this point: Visual representation, cultural landscapes, and the geography of environmental hazards in U.S. National Parks. *GeoHumanities: Space Place, and Humanities.* 6: 314–346.
Youngs, Yolonda, Dave White, and Jill Wodrich. 2008. Transportation Systems as cultural landscapes in national parks: The case of Yosemite. *Society and Natural Resources* 21: 797–811.

CHAPTER 3

From Land Degradation to Habitat Loss: A Qualitative Assessment of Vegetation Cover in Protected Areas of Arid Lands

Xochizeltzin Castaneda-Camacho

Introduction

Land-cover change is a major driver affecting biodiversity and conservation efforts (Bertzky et al. 2019). Major advances in remote sensing have produced insights about changes in forest cover, intensification of croplands, and expansion of urban areas. However, land degradation which is a multifaceted form of land-cover change remains poorly analyzed (Turner II, Lambin, and Verburg 2021). Land degradation is the long-term loss of ecosystem function and services (UNEP 2007) that occurs in one or more of the land resources: vegetation, soils, rocks, water, air, or climate (Yengoh et al. 2014). This work analyzes decline in vegetation cover in

X. Castaneda-Camacho (✉)
Department of Environmental Studies, St. Olaf College, Northfield, MN, USA
e-mail: xochi@stolaf.edu

J. Weber and S. Sultana (eds.), *The Changing Geography of National Parks and Protected Areas*,
https://doi.org/10.1007/978-3-031-74653-6_3

three PAs of arid lands of México to narrowing habitat loss. It is a qualitative approach that shows the mirror-image opposite to what is suitable habitat.

Assessing and monitoring land degradation requires a close-up overview on the affected areas. Decline in vegetation cover, landscape modification, and habitat loss are interconnected processes affected by local pressures and external drivers (Cowie et al. 2018). Numerous studies of vegetation cover change conducted thus far focus on time-series analysis reporting the amount of cover change by using remotely sensed data and vegetation indexes (Priya et al. 2023). The qualitative state of vegetation degradation cover has been largely overlooked, and most importantly, it cannot be stated directly using only remote sensing techniques (Hill et al. 2008). Acknowledging the existent aggregated historical-legacy effects of degradation in vegetation of the PAs under analysis (declared more than 20 years ago), this work employs a single year-time qualitative perspective of habitat loss.

Land degradation processes demand careful attention in fragile ecosystems. Arid lands, including deserts, savannas, and grasslands, comprise nearly one-third of the planet's land mass (Bertzky et al. 2019) and are home to one-quarter of the world's population. Flora, fauna, and human activities are vulnerable to all types of change in these precarious environments.

In arid lands, vegetation degradation increases the risk of species extinction, soil erosion, and desertification. It entails reduction in net primary productivity, declining in species diversity, and degeneration in the nutritional value of plant populations for the faunal biota (UNEP 2007). In sum, unsuitable habitats lack the capacity to provide conditions for wildlife species to meet their needs for resources, shelter, and successful reproduction (Turner and Gardner 2015).

Distinguishing and tracing unsuitable habitat in arid lands is challenging. Landscape changes may be simultaneously nuanced and abrupt. During image processing, the relatively sparse configuration of plants leads to confusion with bare-rocky soil or sand. The spectral signature of human features such as settlement or unpaved roads leads to confusion with those of bare rocks or sand. Disturbed zones may be in a successional cycle (regeneration) holding ephemeral plants that grow spontaneously only after heavy rains. Agricultural lands may be active or inactive. In the former case the spectral signature may resemble vegetation growing on deposition zones. For the latter, parcels may have a colonizer-plant-species cover in the successional cycle. In this work distinguishing the dominant natural vegetation cover is key to understanding habitat loss.

Foundational: Vegetation System

The state of an ecosystem is determined by its primary productivity, the vegetation composition, and the ecosystem services it offers (Yengoh et al. 2014). The primary productivity is the plant biomass per unit area which increases with species diversity through niche complementarity (Guo 2007). The vegetation composition refers to what and how much is present in each cover type (Turner and Gardner 2015). The ecosystem services of vegetation are carbon storage, climate regulation, erosion control, groundwater recharge, wildlife habitats, and human well-being through goods: wild food sources, firewood, raw materials, and medicinal plants.

The configuration of vegetation describes the spatial continuity of cover types across the landscape. Larger patches support a greater number of species since they contain more local variability in microclimate, topography, and structure of plants (Turner and Gardner 2015). In arid lands the growth, density, and configuration of plants are determined by factors that include geomorphology, soils, extreme daily-weather variability, and evapotranspiration that exceeds the annual precipitation. Plants show low growth-regeneration rates which make them vulnerable to natural or human-activities disturbances (Martorell et al. 2015). For example, a drought event decreases biomass (Zhou et al. 2014). Activities such as overgrazing prevent plant regrowth while livestock increases soil compaction.

According to the principle of hysteresis (Yengoh et al. 2014) the state of vegetation declines in response to stressors. Although low-frequency catastrophic events (i.e. fires) help the system to complete its successional cycle (Guo 2007), high levels of disturbance (size or intensity or frequency) cause biodiversity drops because species with low reproduction rates disappear from the community (Safford and Mallek 2010).

Ecological succession is the development of an ecological community structure (niche-based) after perturbations. The specific characteristics of an ecosystem (community members, environment, and perturbation type) define the states, trajectories, and mechanisms of succession (Zhou et al. 2014). The properties of the new cover substantially differ from the former plant community (Chazdon et al. 2016). Indeed, ecosystems with diminished resilience capacity have lower possibilities to recover their original state.

Assessing vegetation cover change requires selecting one of the three stages of the vegetation cycle. Vegetation phenology occurs in three successive stages: dormancy, growth, and senescence (Moulin et al. 1997). The most important resources for plants in the process of photosynthesis are water and sunlight. Yet, regardless of the type of plant, vegetation growth is triggered by precipitation and temperature (Moulin et al. 1997). In arid lands flowering of plants typically arises between 4 and 8 weeks after rain (Sakkir et al. 2015). To assess vegetation from arid lands that vary markedly in type and composition, the growth stage is the most suitable to observe vegetation that is photosynthetically active. Exceptional contrast in vegetation due to environmental factors (altitude, location) needs careful observation prior to determining the growth period of the targeted years.

Building Blocks

Estimations on current environmental change point out the vulnerability of arid lands (Hiroyasu and Hellmann 2019). Climate change models indicate that arid ecosystems in North America will be particularly affected by global warming (Overpeck and Udall 2020). Decreased and more erratic rainfall patterns, a process known as aridification, will result in vegetation loss and soil degradation, a process known as desertification. The reduction of soil will further reduce and change the nature of vegetation cover, thereby affecting wildlife and humans. As large and as fragile as they are, arid lands remain understudied in comparison with temperate and tropical environments. This situation also holds true for PAs that exhibit subtle but constant environmental change processes. Thus, the implementation of restoration projects in PAs is simultaneously a challenge and a priority.

Studies assessing the quality of vegetation cover are few in the literature in comparison to quantitative analysis. Vegetation quality refers to the 'degree to which vegetation at a site resembles native vegetation in the absence of human disturbance' (Gibbons and Freudenberg 2006). Existing studies exhibit different approaches to understanding the response of vegetation in degraded areas. For instance, Jiang et al. (2023) found that planning and restoration in China improved vegetation quality and terrestrial ecosystems; Wang et al. (2024) characterized vegetation quality on primary productivity and extent of coverage for restoration purposes; Martín-Sanz et al. (2024) studied vegetation structure in relation to soil evolution using a qualitative index on abandoned agricultural systems; Yan et al. (2024) analyzed in the Northern Hemisphere of the planet the effect of quality control methods on vegetation analysis, emphasizing the need of high-quality datasets to improve reliability on results.

Measuring the condition (quality) of vegetation is a controversial topic. Biases emerge on current un-defined frameworks, limitations on both the scale-use (landscape level) and the observer. Recently, a vegetation quality assessment (VQA) indicator has been developed by the program of biodiversity monitoring and management (Boyle et al. 2024). The VQA has proved useful for systematizing data collection (observations) at hectares level, however, it might not be efficient in vast territories. Mapping land cover and land use in the growing season for a simultaneous multisite qualitative assessment requires the use of coarse scale, on site observations, satellite imagery, and remote sensing techniques. These are particularly important for large extent (km^2) field sites in remote areas.

The Assessment of vegetation in PAs in arid lands requires consideration of additional facts. The decrees that function as legal instruments to regulate management plans may change through time. Thus, changes in the size of polygons (legal boundaries) need to be traced. Likewise, variation in types of vegetation or rainfall patterns require precise observation. In the latter case, the monthly-annual variation in climatic regimes can be identified from data at weather station level. Cloud-free satellite imagery is preferred to reduce errors in image processing. Dismissing or ignoring these criteria during the analysis may result in misinterpretation. The analysis of the targeted PAs is addressed acknowledging these key facts.

Methods

Land degradation varies across space according to type, scale, time, and extent. Measuring vegetation degradation involves the use of various techniques and empirical expertise. This work analyzes vegetation cover at pixel level and uses remotely sensed data to construct a classification system of land covers in PAs. Most importantly, the quantitative-data transition to a qualitative approach to offer an overview of land degradation

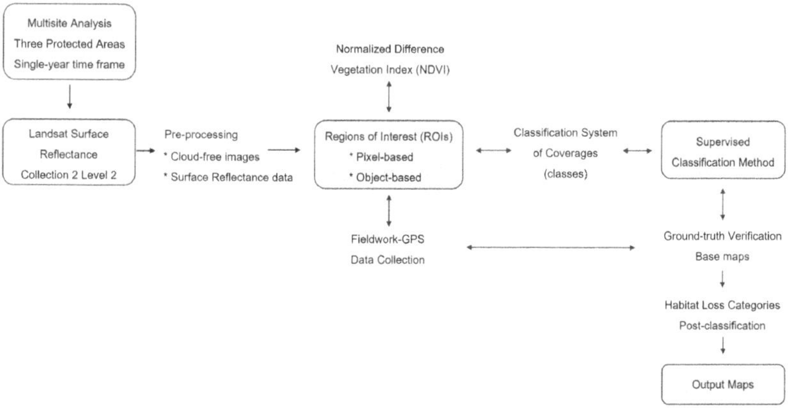

Fig. 3.1 Flowchart of the methodology implemented. *Source* Author

emphasizing five levels of habitat loss. Using the theory of Land System Science and basic concepts of landscape ecology, the main contribution is distinguishing among land covers those that qualitatively met conditions of habitat loss in three PAs that are located far from each other. The methodological design involves elemental criterion and synthesis (Fig. 3.1) providing support for decision-making.

Theoretical Principles

The supervised classification method is suitable for assessing land-cover change through the interpretation of spectral features on remotely sensed data (Jensen 2015). This process requires essential inputs: (1) a representative sampling of the coverages to be examined; (2) an algorithm to distinguish the state of vegetation; (3) an algorithm to discriminate efficiently a specific coverage class among the different classes; and (4) ground-truth validation of results.

The sampling is constructed based on Regions of Interest (ROIs) also named training sites (Lillesand et al. 2015) which are Global-Positional-Systems-locations taken from features in the study area. During the classification process, ROIs provide the key to determine on statistical bases belong probability of each pixel within the different classes of coverages. Thus, accurate location and sample-size of ROIs (during fieldwork and later on the satellite image) are both important. Likewise, it is important to maintain equilibrium in the number of ROIs and adequate spatial distribution per class.

The Normalized Difference Vegetation Index (NDVI) captures bioproductivity trends (Priya et al. 2023) in distinguishing vegetal mass according to its radiometric features (Jiang et al. 2023). The near-infrared and red bands are calculated to show, in a range of values (− 1 to 1), vigorous vegetation (values near or up to 1) in opposition to senescent vegetation (lower values) or other cover types such as water that led to negative values (Chuvieco 2008).

The Maximum Likelihood (ML) algorithm maximizes the probability of a correct classification. According to statistical principles, this classifier assumes that the vectors of each class and spectral band follow a normal distribution (Lillesand et al. 2015). ML computes the mean and covariance of the classes and assign each pixel to a specific class (Campbell and Wynne 2011). The output is a map of the coverages to be ground-truth verified and adjusted.

Study Areas

The study was conducted in three PAs of the Chihuahuan Desert in México (Fig. 3.2). These PAs were selected for their broader significance. They are home to numerous species at risk and align with migratory bird corridors between North and South America. The PAs' geographic location favors the existence of its unique features that make them outstanding hotspots of biodiversity in the world.

In total, the three PAs cover 7394.26 km^2 that comprise nearly the surface of Yellowstone. A major difference, however, is that these PAs are located far from each other. It is a nine-hour drive from the Mapimí Biosphere Reserve in the north, to the Reserve of Real de Guadalcázar in the south, and another four-hour drive from there to the Reserve of the Sacred Site of Huiricuta. Were travel between PAs not difficult enough, travel within the PAs has a special challenge due to their rural settings. Roads are few, predominantly unpaved, sandy (in the case of Mapimí), rough, and often not passable when it rains. Basic services such as electricity and food-supplies are scarce. Gas stations and cell-phone signals are available only near the main roads (before entering the PAs),

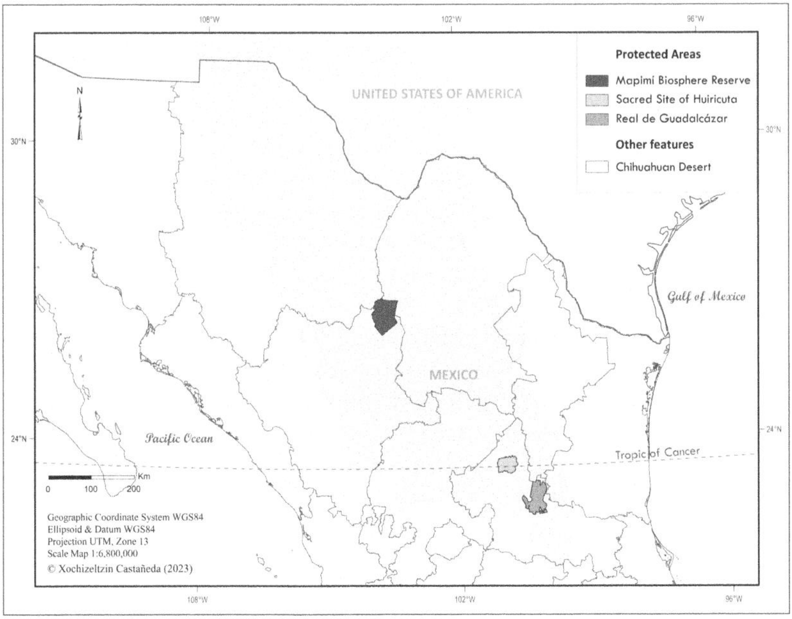

Fig. 3.2 Locations of the study areas regarding political and ecological boundaries. *Source* Author

and surrounding the most populated towns which are few. Other limitations for mobility are imposed by steep terrain, thorny vegetation, snakes, stinging insects, wild carnivores, and extreme weather conditions.

The Federal Mapimí Biosphere Reserve (3423.87 km^2; 103°42′47″ W, 26°39′24″ N, decree-year 1979, updated at 2000), a hot, dry PA is located in the *Bolson de Mapimí* region which belongs to a former lacustric environment. Sedimentary rocks are predominant and few zones at the center-east contain igneous rock. The landscape is dominated by plains and scattered hills (center-east and west) averaging 1300 masl. Growing on dry sandy soils the vegetation cover includes grasslands (Halophiytic-Xerophytic and Halophiytic at west), shrubs (Rossetted and Microphyllous at east and west), and sandy desert plants (north). The *Palomas* Lagoon (north-west) is the most predominant water body in this PA.

The State Reserve of the Sacred Site of Huiricuta (1402.11 km^2, 100°53′47″ W, 23°33′16″ N, decree-year 1994, updated at 2000) holds a territory considered holy by the indigenous *Wixárika* people. This is a mixed semi-arid and temperate region. Sedimentary rocks are prevalent overlaying on surface terrain. The extreme opposite landscape shows the highest-steep mountains at east (peaks up to 3180 masl) and large homogeneous plains to west. Growing in dry-rocky soils, the vegetation includes forest (Oak-Pine and Pinus forests at east ~ 2500 masl; Juniper at north-east), Chaparral (north-east), and shrubs (Microphyllous and Pachycaulous at west; Rossetted scrub at east ~ 2000 masl). The region is mainly drained by non-permanent streams.

The State Reserve of Real de Guadalcázar (2568.28 km^2, 100°19′30″ W, 22°56′32″ N, decree-year 1997, updated in 2006) in a transition zone between the desert (north) and the temperate forest mountain ranges (south) laid on an ancient-tectonic contact area. Multiple fractures on terrain are exhibited among abundant sedimentary rocks. Landscape is dominated by mountains at center-west (highest peaks ~ 2000–2560 masl), south-west, and the north-west edge. Dry-sandy soils correlate with shrubs whereas brown–red fertile soils support forest. The vegetation cover includes various types of shrubs (Rossetted and Microphyllous from center to north, Submountain at south-east), and forest (Oak-Pine associated at south-west). A few small-size water bodies exist in the south but non-permanent streams are predominant in the region.

Data and Sources

The assessment of vegetation cover is based on a Landsat Surface Reflectance Collection 2 Level 2 dataset (USGS 2021) that has been atmospherically corrected (water vapor, ozone, and aerosol optical thickness) through the Land Surface Reflectance Code software (USGS 2020f). This dataset provided by the United States Geological Survey comprises five free-cloud images—carefully selected after rainy season—taken by the Landsat 8 Operational Land Imager (OLI) satellite that uses the Thermal Infrared Sensor (TIRS). This sensor is Tier 1 data, meaning it has been calibrated across the Landsat sensors in addition to having the highest image quality in precision terrain processing. The 30 meters spatial resolution images average 4.86 meters of Root Mean Square Error (RMSE).

Table 3.1 Dataset for the assessment of three protected areas

Protected area name	*Time frame (year)*	*Month-day*	*Scenes*	*GPS-location*	
				2019 y	*2021 y*
Mapimí Biosphere reserve	2020	Oct 10	2	160	491
Sacred Site of Huiricuta	2020	Oct 28	1	92	276
Real de Guadalcázar	2019	Dec 6, 13	2	39	231
Total			5	291	998

GPS global positional system, *y* year

The analysis of the three PAs was supported with data from coarse and fine scale (Table 3.1). The 35 closest weather stations (in total) served as primary input-source (CONAGUA 2021). Topographic maps from the National Institute of Statistics and Geography, or *Instituto Nacional de Estadística y Geografía* (INEGI) covering 7394.26 km^2 let analyzing features in shapefile and at ground level.

The 2020-analysis of the Mapimí Biosphere Reserve was based on two satellite images (USGS 2020a, 2020b, 2020c, 2020d) and eight topographic maps 1:50,000 scale (INEGI 2003, 2017c, 2017d, 2018a, 2018b, 2018c, 2018d, 2018e, 2018f). The 2020-analysis of the State Reserve of the Sacred Site of Huiricuta used one image (USGS 2020e) and six topographic maps 1:50,000 scale (INEGI 2015, 2016a, 2016b, 2016c, 2016d, 2016e). The 2019-analysis of the State Reserve of Real de Guadalcázar was carried out by using two images (USGS 2020a, 2020b) and six topographic maps 1:50,000 scale (INEGI 2016f, 2016g, 2016h, 2016i, 2017a, 2017b).

The Land Use and Vegetation Cover map from México Serie VI (1:250,000 in shapefile, national level) guided the initial classification system of coverages per type (INEGI 2017a). A total of 1289 Global Positional System (GPS) data collected along two summer fieldwork (2019, 2021) supported the identification of coverages in the PAs. The interpretation (digital numbers) and data processing of imagery was done in ENVI software (5.6.1 + IDL 8.8.1 version). The spatial analysis of features in vector format and the cartography was created using the ArcMap software (10.8.2 version).

Procedures

The supervised classification method using the Maximum Likelihood algorithm allowed the identification of coverage types associated to habitat loss. The related procedures were developed in stages. Initial fieldwork conducted in the 2019-summer let gain familiarity with the general features of the PAs. The conditions suitable for research were examined: accessibility, feasibility, and safety. A total of 291 features-locations were registered on site by GPS and photos. The observations pointed out the predominant vegetation types and degradation zones.

The satellite images were processed using the surface reflectance band. The raw data were converted to floating point reflectance (0–1) values using two scale factors: 0.0000275 (gain) + − 0.2 (offset). The assessment was addressed based on the 2019-year for the State Reserve of Real de Guadalcázar, and 2020-year for both the Federal Mapimí Biosphere Reserve and the State Reserve of the Sacred Site of Huiricuta. A total of 291 GPS-locations and the Mexican land use-cover map (1:250,000) served as reference for drawing 1727 (total) ROIs on the scenes. The Normalized Difference Vegetation Index (NDVI) was created for the PAs and served as reference to distinguishing healthy green vegetation and low-density vegetation cover. Three base maps (one for each PA) showing a classification system of coverages were constructed. These maps were ground-truth verified in the 2021-summer fieldwork. A total of 998 locations were reached out to confirm whether the maps accurately portray the coverages of the PAs. The observations were registered on site by GPS and photos.

The base maps were adjusted based on the fieldwork observations (998 GPS-locations). The location-criterion was key for generating a representative sampling for the classification of coverages. The ROIs were allocated (on scenes) using the core (spatially) of the cover type. In a hundred-percent base, the number of ROIs was adjusted to averaging 45 per each cover type (1494 total) ensuring equilibrium for statistical representativeness. The ROIs were re-drawn and re-sized using one-pixel level and consistent spatial distribution. The name of the classes (cover types) was properly adjusted when necessary. Finally, the natural vegetation cover types were mapped in a single category in order to isolate five classes that represent habitat loss. The qualitative level was assigned to each according to the negative effect (extreme-strong).

Validation

The validation of maps from supervised classification typically follows a quantitative approach using an aleatory sampling on the cover types to reveal in an error-matrix table their accuracy in percentage (Mzid et al. 2021). This study uses an empirical guided observation method for corroboration instead of a random test. The reasons are based on the physical characteristics of the PAs: (1) extreme hot-dry weather; (2) location in remote zones; (3) considerable large surface (7394.26 km^2, almost equivalent to the Yellowstone National Park surface); and (4) distance between them. Therefore, 1289 sites (GPS-locations) were visited for ground-truth verification of the cover types.

The initial classification system was constructed on a large number of vegetation cover types for each PA: Mapimí Biosphere Reserve (15), Sacred Site of Huiricuta (18), and Real de Guadalcázar (18). However, preliminary exercises showed that many details in the number and type of classes (coverages) resulted in erroneous classification of features. For example, exposed rock zones were assigned erroneously as settlements. This confusion comes from the 30 meters spatial resolution of the imagery which is considered coarse scale (at pixel level) to discriminate efficiently among features that are typically mixed in a rural environment.

The discrimination of coverages followed key criteria: (1) fewer number of classes improved accuracy in the identification of classes; (2) precise location of sampling (ROIs) improved classification of coverages. The final classification system of each PA contains 10 (Mapimí), 13 (Huiricuta), and 11 (Guadalcázar) classes which is a significant lower number in comparison with the initial classification system (Table 3.2).

Results

The habitat loss categories in maps (Figs. 3.3, 3.4, 3.5) point out zones whose current conditions diminish, prevent, or not provide shelter, space, and suitable characteristics to satisfy wildlife survival. These zones essentially disrupt the connectivity among natural vegetation covers and other natural features from the landscape. The level assigned to the classes (five) from extreme to strong emphasizes in qualitative terms the level of degradation over natural vegetation cover.

Table 3.2 Classification system of coverages in the protected areas: 39 types

	Class type	*M(i)*	*M(f)*	*H(i)*	*H(f)*	*G(i)*	*G(f)*
1	Agricultural and/or barren land				*		*
2	Agriculture-irrigation (annual)	*		*			
3	Agriculture-irrigation (annual/semi-annual)	*					
4	Agriculture-subsistence (annual)			*		*	
5	Agriculture-subsistence (annual/all year)					*	
6	Agriculture-subsistence (annual/semi-annual)			*			
7	Bare surface		*		*		*
8	Chaparral			*	*		
9	Diminished vegetation cover		*		*		*
10	Halophiytic grassland	*	*				
11	Halophiytic grassland SP	*					
12	Halophytic xerophytic	*	*				
13	Halophytic xerophytic SP	*					
14	induced grassland	*		*		*	
15	Human-induced grass and/or vegetation-use for grazing animals				*		*
16	Juniper forest				*		
17	Juniper forest SP			*			
18	Mezquital	*					
19	Mezquital xerophytic					*	
20	Mezquital xerophytic SP					*	
21	Microphyllous desert scrub	*	*	*	*	*	*
22	Microphyllous desert scrub SP	*		*		*	
23	Natural grassland	*					
24	Nude soil	*					
25	Oak forest			*	*	*	*
26	Oak forest SP			*		*	
27	Oak-pine forest			*	*	*	*
28	Oak-pine forest SP			*		*	
29	Pachycaulous scrub			*	*	*	*
30	Pinus forest			*	*		
31	Rossetted desert scrub	*	*	*	*	*	*
32	Rossetted desert scrub SP			*		*	
33	Sandy desert vegetation	*	*				
34	Submountain scrub					*	*
35	Submountain scrub SP					*	
36	Unstable vegetation cover		*				
37	Urban area			*		*	
38	Vegetation in deposition-zone		*				
39	Water	*	*	*	*	*	*

(continued)

Table 3.2 (continued)

Class type	*M(i)*	*M(f)*	*H(i)*	*H(f)*	*G(i)*	*G(f)*
Total	15	10	18	13	18	11

The table shows the initial-final classification of cover types in the Mapimí Biosphere Reserve (M), the Sacred Site of Huiricuta (H), and the Real de Guadalcázar (G). The five classes indicated with numbers 1, 7, 9, 15, and 36 are related to different forms of the habitat loss phenomenon
SP successional phase (revegetation), *i* initial, *f* final

Extreme

The extreme level is represented by the 'bare surface' class which depicts: (1) soils that regardless of the season no longer support the development of plant species; (2) abandoned parcels without crop(s) and no photosynthetic activity; (3) occasionally, bare vegetation cover; (4) small-sparce settlements; (5) exposed rocks (due to human activities or natural processes such as erosion or fractures); and (6) unpaved roads.

Extreme-moderate

The extreme-moderate level is represented by the 'unstable vegetation cover' class which describes zones where native vegetation was cleared but the soils still support the development of plants which typically are in the successional phase. These fragile plants show up only after the annual rainy season. At the stage of regrowth, they provide temporary ecosystem services for wildlife species. The state and 'stability' of this vegetation cover rely on seasonal weather and human influences such as cattle raising.

Severe

The severe level is represented by the 'human-induced grass and/or vegetation-use for grazing animals' class. These are zones where the natural vegetation cover: (1) has been replaced by pastures to raise cattle predominantly large-size; and/or (2) serves for grazing animals such as cows, sheep, and goats. Since animals move freely in the lands, the negative effect of habitat loss expands its boundaries. In arid lands the large-size animals cause the compaction of soils, whereas the small-size animals prevent the regeneration of plants (low-rate growth) by eating

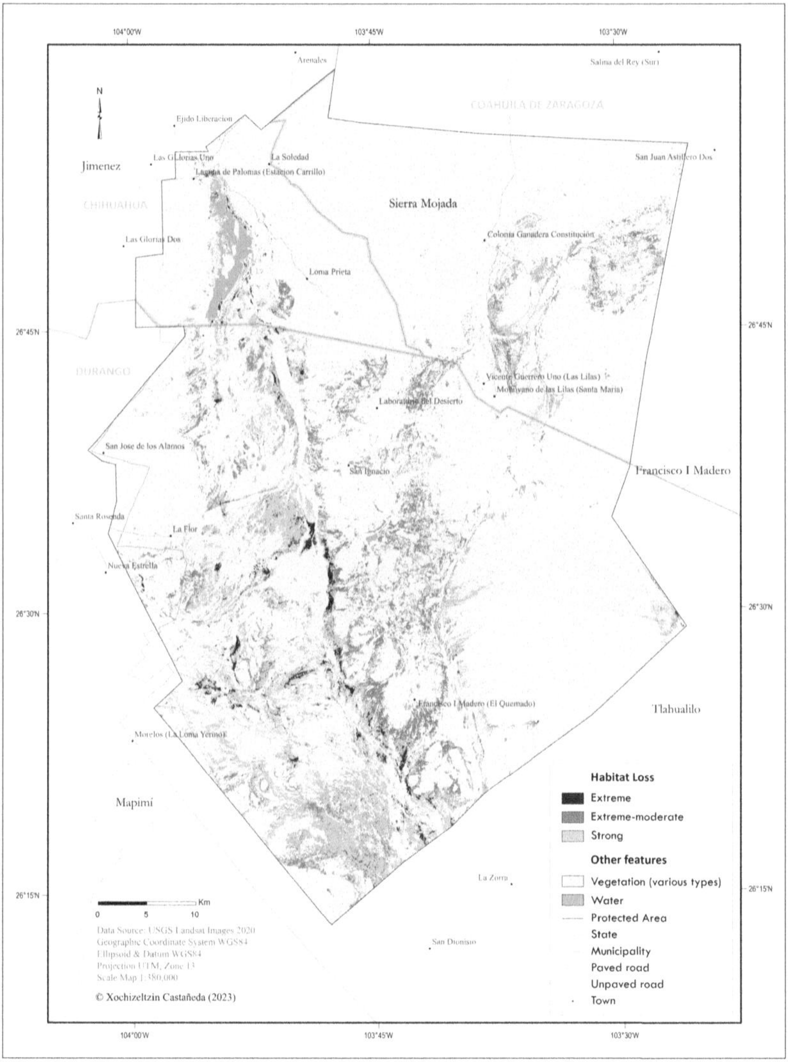

Fig. 3.3 Habitat loss in the Federal Mapimí Biosphere Reserve, 2020. *Source* Author

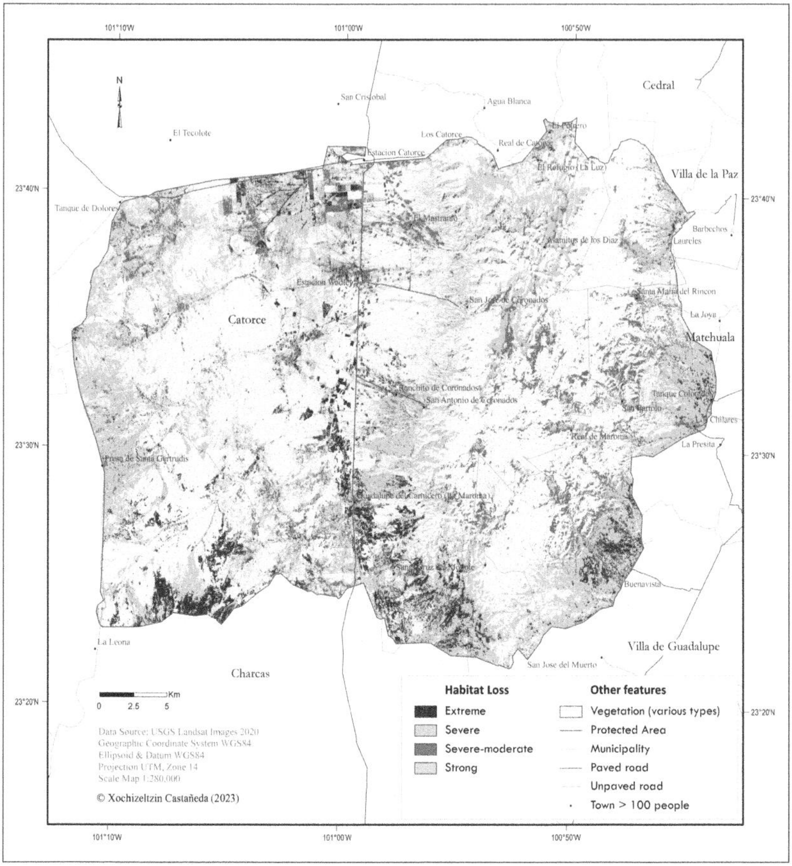

Fig. 3.4 Habitat loss in the State Reserve of the Sacred Site of Huiricuta, 2020. *Source* Author

the tiny regrowths (often preferred over any other stage-cycle of the vegetation).

Severe-moderate

The severe-moderate level is represented by the 'agricultural land and/ or barren land' class. It depicts parcels in active (crops) or non-active

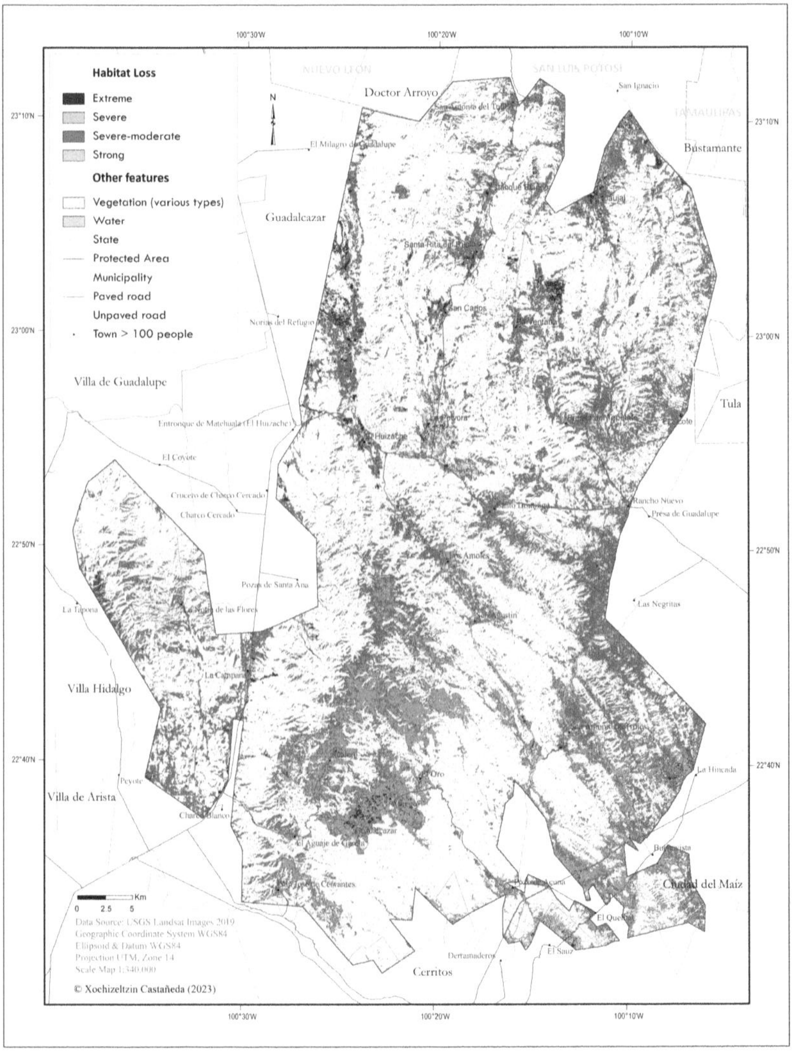

Fig. 3.5 Habitat loss in the Reserve of Real de Guadalcázar, 2019. *Source* Author

(without crops) stages related to smallholder subsistence or mechanized-irrigation systems. The parcels with crops show photosynthetic activity but minimum or none benefits for wild-life. The non-active parcels that have been abandoned four or more years often show vegetation in successional phase (revegetation). Given the rural landscape configuration, this class also depicts rural settlements and barren lands with very few or no vegetation (e.g., steep-rocky terrain, unpaved roads).

Strong

The strong level is represented by the 'diminished vegetation cover' class which depicts zones of fragmented patches and sparse vegetation cover. These zones, typically near to bare surfaces may be: (1) abandoned parcels with vegetation in the successional phase; (2) rural settlement; and (3) home for colonizer species. These zones often represent the 'contention area' that surrounds any of the previous categories. It means that without intervention in the short-mid-term they may transform into bare surface or unstable vegetation cover, thus, increasing the prevalent habitat loss surface.

These results serve to re-think current paradigms on nature conservancy models, global policies, and the function of arid lands and ecosystem services for the planet Earth.

Discussion

The rural landscape of the PAs made difficult distinguishing among covers of natural vegetation and land use. The classification system that resulted in categorizing habitat loss covers required running the supervised classification method multiple times. It was found that a smaller number of classes (coverages) and accurate location of ROIs (training sites) produced a more efficient classification of coverages.

The fieldwork conducted in the summer (2019, 2021) season was critical to identify colonizer species and differences in vegetation-use. For example, in the Federal Mapimí Biosphere Reserve it was found that the *Prosopis glandulosa* species (common-name: *Mesquite*) is a colonizer plant which is drought-tolerant and grows (low-height) in disturbed areas.

The ground-truth verification of base maps and the classification system were critical to identify the coverages that represent habitat loss. An outstanding case was tracing the unstable vegetation in the

Mapimí Biosphere Reserve because plants show up only after rainy season (captured in satellite images) but they were missing during the summer fieldwork.

The scale was constructed on five categories that qualitatively represent habitat loss. The 30 meters spatial resolution data was suitable for assessing the large surface covered but this resolution is coarse to classify in very much detail some human features without overlapping them with natural covers. Some generalizations were made prioritizing the identification of unsuitable habitat, thus, settlement and roads were occasionally assigned in more than one category. The categories of habitat loss were measured in quantitative terms, but the level assigned to each followed a qualitative criterion. Future studies of habitat loss would focus on the identification of spatial patterns of vegetation cover change.

Conclusions

The extreme level category represents the long-term loss of ecosystem function and services. These are critical zones since the soil no longer supports the development of plants (regardless the season) because the layer is very thin, it lacks nutrients or bare rocks exist. Reverse this effect would require active assistance and high-cost investment. As for the extreme-moderate level, the lands are still capable of supporting an early stage of ecological succession, however, the new colonizer species are often different and competitors over native plants. Without intervention through restoration projects native plant species may never complete their full successional cycle.

The severe and severe-moderate levels represent the disruption in the continuity of the landscape which materializes in destruction of habitat. The human activities related to cattle raising and agriculture are affected by burning, haying, irrigation, fertilizing, or chemical treatment. The lands related to human-induced grasslands or by animals released for grazing cause the expansion of habitat loss by compacting soils and/or preventing the regrowth of new plants. Vegetation degradation would be mitigated through sustainable management practices.

The strong level represents unsuitable habitat due to reduction in patch sizes and increased distances between patches which prevent reproduction of plant and animal species. These zones surround bare surfaces (extreme level) meaning that the fragmentation process is prone to increase. If so, these areas would fit into one or more of the previous categories.

The zones classified in strong level still offer potential to recover suitable habitat through proactive regulation, planning, and practices.

Acknowledgements I thank William E. Doolittle at the University of Texas at Austin for his contribution as reviewer of the manuscript. This work was supported by UNAM Posdoctoral Programa (POSDOC)—2024.

Declarations

Conflict of interest Xochizeltzin Castaneda-Camacho has a received research grant from: Donald D. Brand Pre-dissertation Fellowship (2018); E.D. Farmer International Fellowship (2019, 2021); American Geographical Society Council Fellowship (2020); Karl Butzer Excellence Award (2021); Tinker Field Research Grant Summer (2021); Conference of Latin American Geography (2021); C.B. Smith Sr. #1 Centennial Chair in US-Mexico Relations (2021); Robert E. Veselka Endowed Fellowship for Graduate Research Travel (2022).

Permissions The materials used in this paper are copyright-author.

References

Bertzky, Monika, Rebecca C. Brock, Lera Miles, and Valerie Kapos. 2019. Climate change mitigation using terrestrial ecosystems: Options and biodiversity impacts. In *Biodiversity and climate change. Transforming biosphere*, ed. Thomas E. Lovejoy and Lee Hannah, 310–322. Yale: Yale University Press.

Boyle, Bradley L., Warn Franklin, Alison Burton, and Raymond E. Gullison. 2024. Vegetation quality assessment: A sampling based loss-gain accounting framework for native. *Disturbed and Reclaimed Vegetation. Ecological Indicators* 158 (December 2023): 111510. https://doi.org/10.1016/j.ecolind.2023.111510.

Campbell, James B, and Randolph H. Wynne. 2011. Image Classification. In *Introduction to remote sensing*, 5th ed., 335–81. New York: The Guilford Press.

Chazdon, Robin L., Pedro H. S. Brancalion, Lars Laestadius, Aoife Bennett-Curry, Kathleen Buckingham, Chetan Kumar, Julian Moll-Rocek, Ima Célia. Guimarães, and Sarah Jane Wilson. 2016. When is a forest a forest? Forest concepts and definitions in the era of forest and landscape restoration. *Ambio* 45 (5): 538–550. https://doi.org/10.1007/s13280-016-0772-y.

Chuvieco, Emilio. 2008. *Teledetección Ambiental*, 3rd ed. Barcelona: Editorial Ariel, S.A.

CONAGUA. 2021. Climatic Statistical Data-Mexican Government. Data from Weather Stations, Monthly Values. Maximum Temperature; La Soledad

Weather Observation Station; ID 8350. 2021. https://smn.conagua.gob.mx/tools/RESOURCES/Mensuales/chih/00008350.TXT

Cowie, Annette L., Barron J. Orr, Victor M. Castillo-Sanchez, Pamela Chasek, Neville D. Crossman, Alexander Erlewein, Geertrui Louwagie, et al. 2018. Land in balance: The scientific conceptual framework for land degradation neutrality. *Environmental Science and Policy* 79 (1): 25–35.

Gibbons, Philip, and David Freudenberg. 2006. An overview of methods used to assess vegetation condition at the scale of the site. *Ecological Management & Restoration* 7 (S1): S10-17. https://doi.org/10.1111/j.1442-8903.2006.00286.x.

Guo, Qinfeng. 2007. The diversity–biomass–productivity relationships in grassland management and restoration. *Basic and Applied Ecology* 8 (1): 199–208.

Hill, J., M. Stellmes, T. Udelhoven, A. Roder, and S. Sommer. 2008. Mediterranean desertification and land degradation. Mapping related land use change syndromes based on satellite observations. *Global and Planetary Change* 64 (1): 146–157.

Hiroyasu, Elizabeth H.T.., and Jessica J. Hellmann. 2019. Invasive species and climate change. In *Biodiversity and climate change. Transforming biosphere*, ed. Thomas E. Lovejoy and Lee Hannah, 258–269. Yale: Yale University Press.

INEGI. 2003. Topographic Map G13B64, La Flor, 1:50,000 Scale, Durango State, Edition 2003, Shapefile. Instituto Nacional de Estadística y Geografía.

INEGI. 2015. Topographic Map F14A34, Villa de Guadalupe, 1:50,000 Scale, Serial III, San Luis Potosí State, Edition 2015, Shapefile. Instituto Nacional de Estadística y Geografía.

INEGI. 2016a. Topographic Map F14A13, Charco Largo, 1:50,000 Scale, Serial III, San Luis Potosí State, Edition 2017, Shapefile. Instituto Nacional de Estadística y Geografía.

INEGI. 2016b. Topographic Map F14A14, Cedral, 1:50,000 Scale, San Luis Potosí State, Edition 2016, Shapefile. Instituto Nacional de Estadística y Geografía.

INEGI. 2016c. Topographic Map F14A23, Tanque de Dolores, 1:50,000 Scale, Serial III, San Luis Potosí State, Edition 2016, Shapefile. Instituto Nacional de Estadística y Geografía.

INEGI. 2016d. Topographic Map F14A24, Real de Catorce, 1:50,000 Scale, Serial III, San Luis Potosí State, Edition 2016, Shapefile. Instituto Nacional de Estadística y Geografía.

INEGI. 2016e. Topographic Map F14A33, Presa de Santa Gertrudis, 1:50,000 Scale, Serial III, San Luis Potosí State, Edition 2001, Shapefile. Instituto Nacional de Estadística y Geografía.

INEGI. 2016f. Topographic Map F14A45, El Milagro de Guadalupe, 1:50,000 Scale, Serial III, San Luis Potosí State, Edition 2001, Shapefile. Instituto Nacional de Estadística y Geografía.

INEGI. 2016g. Topographic Map F14A46, La Ventana, 1:50,000 Scale, Serial III, San Luis Potosí State, Edition 2001, Shapefile. Instituto Nacional de Estadística y Geografía.

INEGI. 2016h. Topographic Map F14A55, Pozas de Santa Ana, 1:50,000 Scale, Serial III, San Luis Potosí State, Edition 2001, Shapefile. Instituto Nacional de Estadística y Geografía.

INEGI. 2016i. Topographic Map F14A56, Lázaro Cárdenas, 1:50,000 Scale, San Luis Potosí State, Edition 2001, Shapefile. Instituto Nacional de Estadística y Geografía.

INEGI. 2017a. Land Use and Vegetation Cover of Mexico Map, 1:250,000, Series VI. Mexico: Instituto Nacional de Estadística y Geografía.

INEGI. 2017b. Topographic Map F14A65, Guadalcázar, 1:50,000 Scale, Serial III, San Luis Potosí State, Edition 2017, Shapefile. Instituto Nacional de Estadística y Geografía.

INEGI. 2017c. Topographic Map F14A66, Buenavista, 1:50,000 Scale, Serial III, San Luis Potosí State, Edition 2017, Shapefile. Instituto Nacional de Estadística y Geografía.

INEGI. 2017d. Topographic Map G13B84, Cartagena, 1:50,000 Scale, Serial III, Durango State, Edition 2004, Shapefile. Instituto Nacional de Estadística y Geografía.

INEGI. 2018a. Topographic Map G13B54, Laguna de Palomas, 1:50,000 Scale, Serial III, Chihuahua State, Edition 2003, Shapefile. Instituto Nacional de Estadística y Geografía.

INEGI. 2018b. Topographic Map G13B55, El Cinco, 1:50,000 Scale, Serial III, Coahuila State, Edition 1973, Shapefile. Instituto Nacional de Estadística y Geografía.

INEGI. 2018c. Topographic Map G13B63, Ceballos, 1:50,000 Scale, Serial III, Durango State, Edition 2002, Shapefile. Instituto Nacional de Estadística y Geografía.

INEGI. 2018d. Topographic Map G13B65, Sierra El Zapatero, 1:50,000 Scale, Serial III, Coahuila State, Edition 2006, Shapefile. Instituto Nacional de Estadística y Geografía.

INEGI. 2018e. Topographic Map G13B74, Las Tortugas, 1:50,000 Scale, Serial III, Durango State, Edition 2004, Shapefile. Instituto Nacional de Estadística y Geografía.

INEGI. 2018f. Topographic Map G13B75, Córdoba, 1:50,000 Scale, Serial III, Durango State, Edition 1973, Shapefile. Instituto Nacional de Estadística y Geografía.

Jensen, John Richards. 2015. Thematic information extraction: Pattern recognition. In *Introductory digital image processing: A remote sensing perspective*, 4th ed., ed. Christian Botting, 361–428. Glenview, IL: Pearson Education Inc.

Jiang, Luguang, Ye. Liu, and Xu. Haixia. 2023. Variation in vegetation quality of terrestrial ecosystems in China: Coupling analysis based on remote sensing and typical stations monitoring data. *Remote Sensing*. https://doi.org/10.3390/rs15092276.

Lillesand, Thomas, Ralph W. Kiefer, and Jonathan Chipman. 2015. Digital image analysis. In *Remote Sensing and Image Interpretation*, 7th ed., 485–608. New York: Wiley.

Martín-Sanz, Juan Pedro, Inmaculada Valverde-Asenjo, Miguel Ángel Casermeiro, José Antonio. Molina, and José Ramón. Quintana. 2024. A new quality index based on soil-vegetation networks to determine the recovery of functionality in abandoned agrosystems. *Ecological Indicators*. https://doi.org/10.1016/j.ecolind.2023.111466.

Martorell, Carlos, Delfín M. Montañana, Carolina Ureta, and María C. Mandujano. 2015. Assessing the importance of multiple threats to an endangered globose cactus in Mexico: Cattle grazing, looting and climate change. *Biological Conservation* 181: 73–81. https://doi.org/10.1016/j.biocon.2014.10.035.

Moulin, S., L. Kergoat, N. Viovy, and G. Dedieu. 1997. Global-scale assessment of vegetation phenology using NOAA/AVHRR satellite measurements. *Journal of Climate* 10: 1154–1170.

Mzid, Nada, Stefano Pignatti, Wenjiang Huang, and Raffaele Casa. 2021. An analysis of bare soil occurrence in arable croplands for remote sensing topsoils applications. *Remote Sensing* 13 (474): 2–24. https://doi.org/10.3390/rs13030474.

Overpeck, Jonathan T., and Bradley Udall. 2020. Climate change and the aridification of north America. *Commentary* 117 (22): 11856–11858.

Priya, M.V., R. Kalpana, S. Pazhanivelan, R. Kumaraperumal, K.P. Ragunath, G. Vanitha, Ashmitha Nihar, P.J. Prajesh, and V. Vasumathi. 2023. Monitoring vegetation dynamics using multi-temporal normalized difference vegetation index (NDVI) and enhanced vegetation index (EVI) images of Tamil Nadu. *Journal of Applied and Natural Science* 15 (3): 1170–1177. https://doi.org/10.31018/jans.v15i3.4803.

Safford, Hugh D., and Chris R. Mallek. 2010. Disturbance and diversity in low-productivity ecosystems. In *Serpentine: The evolution and ecology of a model system*, ed. Susan Harrison and Nishanta Rajakaruna, 249–274. London: The Regents of the University of California.

Sakkir, Sabitha, Junid N. Shan, Abdul J. Cheruth, and Maher Kabshawi. 2015. Phenology of desert plants from an arid gravel plain in eastern United Arab Emirates. *Journal of Arid Land* 7 (1): 54–62. https://doi.org/10.1007/s40333-014-0036-2.

Turner, B.L., II., Eric F. Lambin, and Peter H. Verburg. 2021. From land-use/land-cover to land system science. *Ambio* 50 (1): 1291–1294.

Turner, Monica G., and Robert H. Gardner. 2015. *Landscape ecology in theory and practice*. Second: Springer.
UNEP. 2007. Land. Nairobi, Kenya: United Nations Environmental Programme (UNEP).
USGS. 2020a. Landsat 8, OLI TIRS, December 13, 2019, Path 28, Row 44, UTM Zone 14, WGS84, 17:11 Hrs., RMSE 5.306, 30 m Spatial Resolution, C2L2SP, Tier 1. U.S. Geological Survey Earth Resources Observation and Science Center.
USGS. 2020b. Landsat 8, OLI TIRS, December 6, 2019, Path 27, Row 44, UTM Zone 14, WGS84, 17:05 Hrs., RMSE 4.563, 30 m Spatial Resolution, C2L2SP, Tier 1. U.S. Geological Survey Earth Resources Observation and Science Center.
USGS. 2020c. Landsat 8, OLI TIRS, October 10, 2020, Path 30, Row 41, UTM Zone 13, WGS84, 17:22 Hrs., RMSE 4.322, 30 m Spatial Resolution, C2L2SP, Tier 1. U.S. Geological Survey Earth Resources Observation and Science Center.
USGS. 2020d. Landsat 8, OLI TIRS, October 10, 2020, Path 30, Row 42, UTM Zone 13, WGS84, 17:22 Hrs., RMSE 4.746, 30 m Spatial Resolution, C2L2SP, Tier 1. U.S. Geological Survey Earth Resources Observation and Science Center.
USGS. 2020e. Landsat 8, OLI TIRS, October 28, 2020, Path 28, Row 44, UTM Zone 14, WGS84, 17:11 Hrs., RMSE 5.347, 30 m Spatial Resolution, C2L2SP, Tier 1. U.S. Geological Survey Earth Resources Observation and Science Center.
USGS. 2020f. Landsat 8 Collection 2 (C2) Level 2 Science Product (L2SP) Guide, Version 2.0, September.
USGS. 2021. Landsat Science Products Collection 2 Surface Reflectance Level 2. Landsat Science Products Courtesy of the U.S. Geological Survey Earth Resources Observation and Science Center. 2021. https://www.usgs.gov/core-science-systems/nli/landsat/landsat-collection-2-surface-reflectance
Wang, Lixia, Jixi Gao, Wenming Shen, Mingyong Cai, Yalin Chen, Tong Xiao, Xinsheng Zhang, and Wenfei Tai. 2024. Vegetation quality and its change in the three north shelterbelt forest region in China. *Arid Land Research and Management* 38 (2): 285–298. https://doi.org/10.1080/15324982.2023.2269883.
Yan, Kai, Xingjian Zhang, Rui Peng, Si. Gao, and Jinxiu Liu. 2024. The impact of quality control methods on vegetation monitoring using MODIS FPAR time series. *Forests* 15 (3): 1–17. https://doi.org/10.3390/f15030553.
Yengoh, Genesis T., David Dent, Lennart Olsson, Anna E. Tengberg, and Compton J. Tucker. 2014. The use of the normalized difference vegetation index (NDVI) to assess land degradation at multiple scales: A review of the

current status, future trends, and practical considerations. *Land University Centre for Sustainability Studies* 2014: 1.

Zhou, Jizhong, Ye. Deng, Ping Zhang, Kai Xue, Yuting Liang, Joy D. Van-Nostrand, Yunfeng Yang, et al. 2014. Stochasticity, succession, and environmental perturbations in a fluidic ecosystem. *PNAS* 111 (9): E836–E845. https://doi.org/10.1073/pnas.1324044111.

CHAPTER 4

The Role of National Parks in the Development of Disadvantaged Karst Areas in Some European Countries

Tamás Telbisz and László Mari

Introduction

National parks (NPs) play a complex role in the life of areas where they were established. In addition to their nature conservation tasks, their impact is also significant from a social point of view. Karst areas represent a remarkable proportion of national parks in Europe. Karstification is primarily a natural phenomenon with specific geomorphological and hydrogeological characteristics. However, as a consequence of these natural characteristics, karst regions are often socially disadvantaged areas in many European countries. The above facts logically lead to the main question of our investigations, whether national parks can have a positive effect on the socio-economic development of karst areas.

T. Telbisz (✉) · L. Mari
Department of Physical Geography, Eötvös Loránd University, Budapest, Hungary
e-mail: tamas.telbisz@ttk.elte.hu

J. Weber and S. Sultana (eds.), *The Changing Geography of National Parks and Protected Areas*,
https://doi.org/10.1007/978-3-031-74653-6_4

Below, the elements of the above logical chain are examined. First, the special characteristics of karst regions are reviewed. Second, we present how the number of national parks and geoparks in Europe increased after the Second World War, and the proportion of karst regions within national parks and geoparks is also calculated. Since the concept of geotourism has become more and more important in recent decades (Dowling 2011; Farsani et al. 2011), the relationship between geoparks and karsts is also briefly addressed. Thereafter, the potential impact of karst natural factors on certain social phenomena is taken into account. Next, primarily on the basis of demographic indicators, it is evident that in many European karst areas, social disadvantages can indeed be detected in terms of depopulation and ageing. Finally, we turn to the most important question: the role that national parks play in this context. As a background, it is also presented how the priorities (or tasks) of national parks have changed over the past century and a half. Our own investigations are based on case studies carried out in several national parks in Eastern Central and Southern Europe (Aggtelek National Park/Hungary, Apuseni Nature Park/Romania, Krka National Park/Croatia, Slovak Karst National Park/Slovakia, Tara National Park/Serbia). The relationships between the elements of the park-people-tourist triangle are the focus of these studies. Our research was carried out in the framework of an international project supported by the National Research, Development and Innovation Office Hungary (https://karst.elte.hu/knp/).

Theoretical Background

Examining the relationship between natural settings and socio-economic development is one of the basic questions of geography with several approaches (geographic determinism, nihilism, possibilism). Our idea is in line with the concept of *geographic possibilism*, according to which natural conditions set certain constraints on socio-economic development, but taking into consideration these limits, the internal processes of society drive development (Mercier 2009).

Special Settings of Karst Terrains

Karst terrains are areas where the solubility of the rock (mostly limestone or dolomite) determines the landforms and the way of water drainage. In karst areas, surface watercourses and valleys are almost absent (or at least present in smaller density), thus the topography is characterized by steep mountainsides and plateaus dotted with sinkholes. Infiltration through sinkholes and stream sinks usually plays a decisive role in karst drainage. The water reaches the springs below the surface through caves, and a significant part of the water ends up in underground reservoirs for a long time, which is a very important source of drinking water in many areas of the Earth. Due to the rapid infiltration, the karst soils often have a moisture deficit and thus generally shallower soils develop on karst terrains than on other bedrock under the same climatic conditions. Based on the aforementioned natural conditions, it is possible to outline the effects this exerts on society. Since the availability of water is limited on the karst plateaus (where water collection is only possible via cisterns), they are limited in terms of human settlement. On the other hand, karst springs with abundant water offer good opportunities for settlement mainly at the mountain foot areas. Due to the poor and thin nature of karst soils, the ploughing opportunities are less favourable, therefore karst areas are more suitable for animal husbandry and the preservation of forests. Steep, rocky mountain slopes put an obstacle for traffic. Taking into account all of the above factors, it is not surprising that karst regions are generally relatively sparsely populated areas (Telbisz et al. 2014), though exceptions are also known (Goldscheider et al. 2020). Since population density is lower, natural vegetation could be better preserved in many karst areas. In addition, there exist special calciphil plants. As a result, karst areas often have high biodiversity (Clements et al. 2006; Gorjanc et al. 2022). Furthermore, the special landforms (dolines, collapse sinkholes, caves, and gorges) of karst terrains attract lots of tourists. In total, all of the above factors may play a role in the establishment of protected areas in karst regions, and may contribute to the boom in nature-based tourism in these areas. Figure 4.1 summarizes the relationships mentioned above.

Depopulation of Karst Terrains

In recent years, a number of articles have been published, which quantitatively proved that the population density is not only low in the area of many European karst regions, but in the last half to a century the population has mostly been decreasing. Demographic studies of karst terrains are known from the Velebit Mountains (Croatia; Pejnović and Husanović-Pejnović 2008), the Gömör-Torna-karst (Hungary, Slovakia, Telbisz et al.

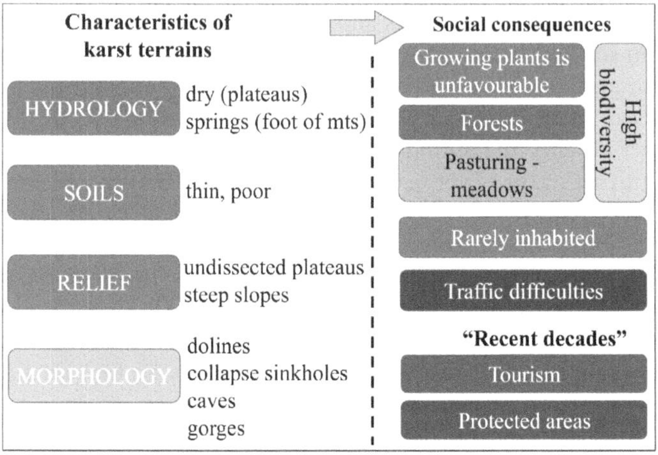

Fig. 4.1 Impact of karst on society

2015), the Apuseni Mountains (Romania, Telbisz et al. 2016), the Tara Mountains (Serbia, Telbisz et al. 2020b), the Tymphi Mountain (Greece, Telbisz et al. 2019), and the wider environment of the Krka National Park (Croatia, Telbisz et al. 2022c). In order to quantitatively characterize the influence of karst on demography, karst regions were compared to their surrounding non-karst areas. Study areas were selected so that most other factors (such as topography, climate, distance from population centres, etc.) were similar to the selected karst and the surrounding non-karst areas. This way, the differences can be traced back to the karstic natural factors. Nevertheless, considering the diversity of geographical factors, it is never possible to have a difference in only one factor, but the above studies approximated this selection strategy as much as possible. Therefore, the mentioned papers have convincingly demonstrated that karstification generally increases the degree of depopulation.

Naturally, the process of depopulation must be interpreted in a broader context, too. In the broadest sense, rural depopulation should be mentioned (Newsham and Rowe 2023). However, by further narrowing the circle, it is unambiguous that the depopulation of the mountainous regions is faster (Milošević et al. 2010, 2011; Kohler et al. 2017). Nonetheless, the above examples demonstrate that the depopulation of karst areas is even faster within the group of mountain settlements. Generally, the direct causes of depopulation are lower incomes in agriculture, lack of employment, weaker infrastructure (roads, electricity, internet), and fewer social institutions (schools, shops, entertainment, etc.). All of this leads to depopulation and the ageing of society. These processes cannot be stopped, but the benefits associated with protected areas can somewhat alleviate these problems (Grau and Aide 2007; Gretter et al. 2017).

Protected Areas

Protected areas are one of the most important means of nature conservation for the benefit of the whole society (Dudley and Stolton 2010; Watson et al. 2014). However, the effectiveness of nature conservation in protected areas also depends on the extent to which different social actors come together and to what extent they support nature conservation goals (Jamal and Stronza 2009). Therefore, examining the relationships in the park–local residents–tourists triangle is a very important topic. In these relationships, conflicts often arise between social actors (Zachrisson et al. 2006). The way to resolve conflicts is through getting to know each other's interests. In the current research, these relationships are examined through the example of national parks (or, in one case, a nature park) that were established in karst areas.

Before zooming in on the present situation, it is worth examining how the system of goals of national parks has changed since the foundation of the first national park (Yellowstone National Park founded in 1872). In addition to the preservation of "wild nature" and the protection of endangered species, the original goals of the first national parks in the USA and its followers also included the promotion of tourism and the expression of national identity (Frost and Hall 2015). Much later, in the twentieth century, when the idea of the "national park" arrived in Europe, there were already more densely populated regions and much less "wilderness areas", thus the idea arose that "cultural landscapes" should also be protected, and that the national park should also provide opportunities for recreation. As a result of new scientific knowledge in the field of ecology in the second half of the twentieth century, the original aim to preserve individual species was enlarged to preserve biodiversity and whole ecosystems. It was also recognized that national parks (or protected areas in general) are excellent locations for education. Moreover, the special role of national parks in terms of scientific research was also emphasized. It was relatively lately, from the 2000s, that the approach that protected areas should also contribute to the promotion of regional development became highlighted (Butler and Boyd 2000; Mose 2007). Today, all of these goals are formulated in the IUCN (International Union for Conservation of Nature) recommendations for protected areas (Leroux et al. 2010). For instance, it is written that national parks "provide a foundation for environmentally and culturally compatible spiritual, scientific, education, recreational and visitor opportunities" (Leroux et al. 2010, p. 610), and "national parks are managed in a way that may contribute to local economies" (Pullin et al. 2013, p. 12).

Several differences can be mentioned between European and US national parks, which were presented by Frost and Hall (2015). Here we only want to shed light on a slight but important technical difference, which may make the situation of the national parks presented below more understandable for American readers. National parks in the USA are usually much larger in area, but they are generally accessible by car from only a few directions that makes it possible to collect entrance fees from visitors. In contrast, the majority of European national parks are usually smaller in area, more densely populated with settlements, and accessible from several directions, therefore it is rare to find a national park where an entrance fee is charged for the entire park. Instead, entrance fees are rather linked to specific sights and locations (in the case of karsts, mainly caves and occasionally gorges or waterfalls are the attractions that can be viewed with a ticket).

Research Methods

Our methodology consists of the following elements.

Based on various databases, it was examined how the number and spatial distribution of national parks and geoparks increased in European countries. Further on, national parks and geoparks were classified according to the proportion of karst areas in them. Three categories were applied: not karstic, partly karstic, and completely (or mostly) karstic.

The further results presented below are based on questionnaire surveys and semi-structured interviews conducted in four national parks and one nature park. Questionnaire surveys are typical elements of studies dealing with the role and development opportunities of national parks (Trakolis 2001; Papageorgiou and Kassioumis 2005; Zurc and Udovč 2009; Šulc and Valjak 2012; Zgłobicki and Baran-Zgłobicka 2013; Krpina 2015; Mika et al. 2019; Zawilińska 2020). In the present research, questionnaire surveys were conducted in the examined parks to know the opinions of the local residents and visitors. The questionnaires were basically similar in each park, though small changes were made to adapt the questions to the special situation of each park. There were 28 questions for visitors and 34 questions for local people, mostly with multiple-choice or Likert-scale questions, but there were also some open-ended questions. Some sample questions (non-exhaustive) from the questionnaire prepared for local people were:

- What are the values of this landscape?
- How do you see the economic situation of your settlement?
- How often do tourists come to your settlement?
- What are the advantages/drawbacks due to the national park for local people?
- How important should the preservation of geological (biological, etc.) values be in the national park?

Some sample questions (non-exhaustive) from the questionnaire prepared for tourists were:

- What are the values of this landscape?
- How did you get information about this national park?
- Was it important for you that this area is a "National Park"?
- What kind of tourist development would you think is important for this national park?
- Have you ever heard the expression: "geotourism"? If yes, please explain it shortly.
- How important should the preservation of geological (biological, etc.) values be in the national park?

Filling the questionnaire was typically a few minutes in most cases. For the visitors, the questionnaire was available in several languages. The questionnaire surveys were conducted between 2018 and 2021. Questionnaires were mostly filled onsite (except Krka, where online filling was also available for local people), with direct, face-to-face questioning, with the help of assistants. The method of convenience sampling was used in all cases. The results were evaluated in terms of descriptive statistics. Finally, we conducted long interviews with prominent actors, such as the park managers, external experts, and the mayors of the settlements in and around the studied parks. The main topics of the interviews were: how park managers interpret the functions of the park; what conflicts they perceive; and how the park contributes to the socio-economic development of the region. Their answers were qualitatively evaluated.

Study Areas

The analysis using the first method described above was carried out in a European context. The other methods were applied to selected national (or nature) parks at the following locations: Aggtelek National Park (Hungary), Apuseni Nature Park (Romania), Krka National Park (Croatia), Slovak Karst National Park (Slovakia), and Tara National Park (Serbia). The locations are shown in Fig. 4.2, and some photos about typical karst features of these parks are presented in Fig. 4.3.

The *Aggtelek National Park* (Hungary, 199 km^2) was founded in 1985, mainly because of the caves found here, of which the 25 km long

Fig. 4.2 Studied national (nature) parks in Eastern Central and Southern Europe

Fig. 4.3 Photos from the studied national (nature) parks. **a** Aggtelek NP (Dragon head dripstone in the Baradla Cave); **b** Slovak Karst NP (gorge of Zádielska dolina); **c** Apuseni NP (sinkholes on Padis Plateau); **d** Apuseni NP (gorge and waterfall in Galbena valley); **e** Krka NP (Skradinski Buk waterfalls); **f** Tara NP (Perućac lake and Tara plateau). Photos: Telbisz, T. (**a**–**c**, **e**, **f**) and Imecs Z. (**d**)

Baradla-Domica system (partially extending into Slovakia) is the most famous. This cave is the focus of tourism to this national park (Telbisz et al. 2020a, 2022b). From a natural point of view, the Aggtelek Karst forms a connected landscape with the Slovak Karst (Slovakia), where the *Slovak Karst National Park* (346 km^2) was founded in 2002. The cave systems of the two areas altogether constitute a UNESCO World Heritage Site. The landscape is characterized by karst plateaus and gorges, and the typical land cover is forest. From a social point of view, both areas are relatively poor, rural areas, although there are also some nearby towns on the Slovak side.

Unlike the other investigated areas, the *Apuseni Nature Park* (Romania, 758 km^2) is "only" a nature park (IUCN category V), but the same surveys were conducted in this park as well (Imecs et al. 2022). The area is characterized by extremely spectacular karst formations (caves, collapse sinkholes, gorges). In addition, the bioheritage of the park is also very significant as the area is largely covered by forests. There are 53 small settlements in the area of the nature park, with about 10,000 people living there. Their main source of income are forestry and agriculture, but recently the role of tourism has also increased in terms of income generation. The area was declared a nature park in 2000.

The *Krka National Park* (Croatia, 109 km^2) is different from the other parks of this study, as it is located close to the coast, and it is a low hilly area with little forest cover. This national park was founded in 1985. Its main attractions are the limestone tufa waterfalls and an abundance of water. Partly due to the proximity to the coast, where mass tourism is present, and partly due to its spectacular forms, visitation to this national park is extremely high (with more than 1 million people per year) compared to the other examined parks (2–500 thousand people per year). Nonetheless, the inland area is relatively sparsely populated and has a rural character (Telbisz et al. 2022a, c).

The *Tara National Park* (Serbia, 183 km^2) is a karst area in medium mountains with plateaus and gorges, similar to the previous examples (Telbisz et al. 2020b, 2021). Despite the fact that it is a karst area, tourist caves are not found here. A large part of the area is covered by forests, which are directly in the management of the national park, and the bioheritage of the park is also very significant. There are small settlements in and around the park, as well as a small town nearby. This national park was founded in 1981, making it the oldest of the examined parks.

Results

Karstic National Parks and Geoparks in Europe

Based on the analysis of European national parks and geoparks, we found that there are currently 106 partially or mostly karstic national parks in Europe, which is 23% of all national parks. As for geoparks, 49% of them are partially or mostly karstic (Telbisz and Mari 2020). Therefore, it is concluded that within protected areas in general, but especially in terms of geoheritage, the karst areas are highly significant. Their spatial distribution within Europe naturally corresponds to the spatial distribution of karst terrains, i.e. the British Isles as well as Central and Southern Europe have a significant number of partially or mostly karstic national parks or geoparks (see Fig. 4.4 and our online map at http://karst.elte.hu/maps/EuKarstMap.html). Regarding the establishment of national parks, we can see a gradual acceleration. From 1950 to 1990, an average of 4.2 national parks were established annually in Europe, and from 1990 to 2010, the growth rate was 11.5 national parks per year on average. In the most recent decade, however, the rate has already slowed down significantly (Telbisz and Mari 2020).

Results Drawn from Questionnaire Surveys and Interviews

The following results are based mostly on our questionnaire surveys with local residents and visitors, as well as the semi-structured interviews with local stakeholders and experts. Altogether, 25 interviews were conducted, and 2891 questionnaires were filled: 1930 questionnaires by tourists and 961 questionnaires by local residents in the five studied parks. In this chapter, the main findings are presented, but the detailed results of the surveys are found in the following articles: Telbisz et al. 2020a, b, 2021, 2022a, b; Nestorová Dická et al. 2020; Brankov et al. 2022; Imecs et al. 2022, Kőszegi et al. 2022, Kovačević-Majkić et al. 2022.

Table 4.1 highlights some important generalized results, and it qualitatively compares how significant are the different processes (or factors) in the studied national (or nature) parks. According to our analysis, depopulation is characteristic of all the examined karst regions without exception, and even the establishment of the national park could not really stop this process. From a demographic point of view, in most places only a few settlements show demographically positive signs, these are the settlements that benefit most directly from the positive social

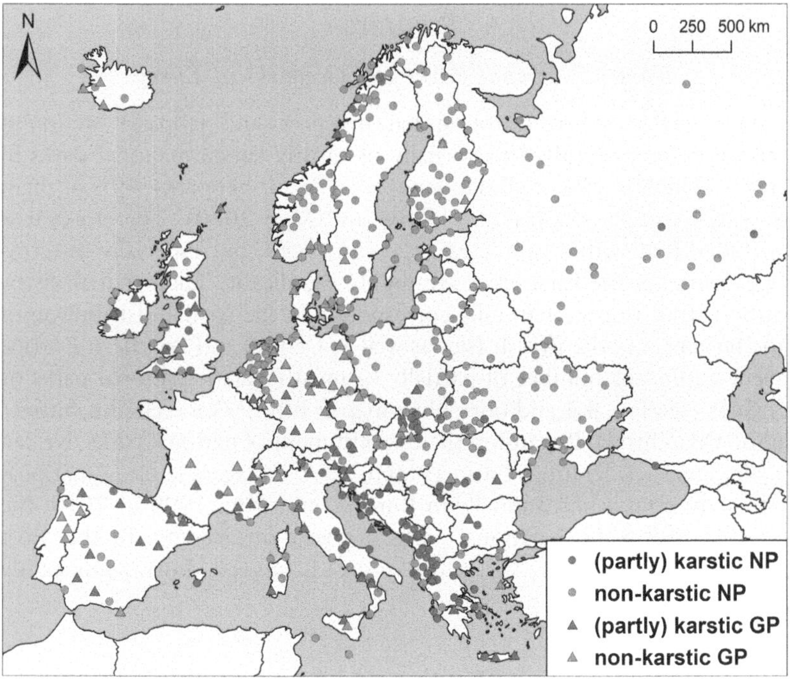

Fig. 4.4 Locations of national parks (NP) and geoparks (GP) in Europe classified as (partly) karstic or not karstic

effects of the national park (through job opportunities or participation in hospitality). The infrastructure (e.g. road network) is in an improving condition everywhere, and in comparison to the surroundings of these parks, the infrastructure is generally of better quality, which is due to the fact that the opportunities to apply for funding are better in case of the national (nature) parks than in the neighbouring settlements, therefore more financial support arrives to the park areas. In terms of job opportunities, we can talk about direct job opportunities (when the national park is the employer) and indirect opportunities (which can be linked mostly to tourism). Among the examined national parks, the Krka NP is the one that provides direct employment to the most people, while the number of jobs directly connected to the park is negligible in the case of Slovak Karst NP and Apuseni NP. In the case of Tara NP and Aggtelek

NP, we can speak of a medium value, and in the case of Aggtelek NP, it is emphasized that the national park is the largest employer in the subregion. It is also interesting to examine which sector provides most of the national park's revenue. In the case of Krka NP, where the annual number of visitors is the highest (currently well over 1 million), tourism is of course the main source of income. In the case of Tara NP, interestingly, the sector with the most revenue is forestry, which takes into account aspects of nature conservation, yet it can operate with significant economic benefit using the territory of the national park. In the other cases, the state budget support is the most important item. Finally, the investigated national parks were categorized in terms of the importance of tourism, the internationality of visitors, and the type of main natural attractions (see Table 4.1).

By examining the different goals of the stakeholders, the following conclusions are drawn. In most cases, it can be observed that the state budgets are in a "difficult situation" everywhere, i.e. governments are trying to reduce their costs spent on nature conservation, and specifically on the maintenance of national parks. Therefore, the expectation is communicated to the park managers that if they want to maintain the operation of the park at the same level (including the continued

Table 4.1 Importance of different factors in the studied national (nature) parks

Factor	*Aggtelek NP*	*Apuseni nature park*	*Krka NP*	*Slovak karst NP*	*Tara NP*
Depopulation	depopulation is typical in each of them—not stopped by the presence of the NP except few developing settlements closest to the main attractions				
Infrastructure (roads)	Improving in each case (but still not perfect everywhere)				
Job possibilities	Medium	Few	Many	Few	Medium
Main source of income of the NP	State	State	Tourism	State	Forestry
Importance of tourism	Medium	Medium	Very high	Medium	Medium
Internationality	Domestic	Domestic	International	Domestic	Domestic
Natural attractions	Caves, forests	Caves, forests, gorges	Waterfalls, lakes	Caves, forests, gorges	Forests, gorges

employment of local people), they should try to get as much independent income as possible through tourism and tenders. Despite the above difficulty, however, it is undoubtedly a fact that significant resources are available for the development of the infrastructure (partly through state and partly through EU funds). The managers of the national parks are mostly committed to nature conservation, but the current financial situation poses serious limitations for them. Their primary task is to achieve nature conservation goals, and they would like to leave the business to local entrepreneurs, but partly due to historical reasons (namely, the former socialist regime), the local entrepreneurial layer is weak in many cases. In most places, local residents would be happy to welcome more tourists than at present, as they hope that this will lead to new job opportunities and more income (Fig. 4.5). Also, in most places they agree (although not everywhere) that the national park should take a significant part in the management of tourism. On the other hand, they do not like certain restrictions related to protected areas (such as restrictions on construction, certain agricultural activities, and forest gatherings).

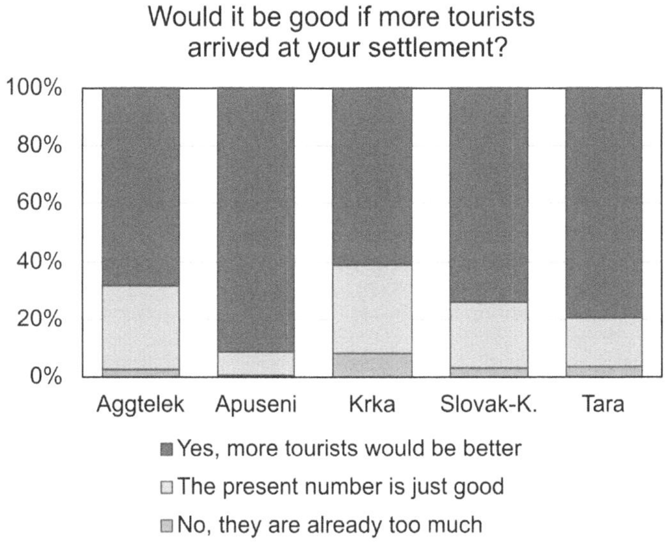

Fig. 4.5 Survey answers of local residents to the question "Would it be good if more tourists arrived at your settlement?"

Overall, weighing the advantages and drawbacks, the surveys show that the majority of local residents experience the advantages of the national park more strongly than its drawbacks (Fig. 4.6). The exception to this trend is the Apuseni NP, where the dialogue between the various stakeholders is presumably weaker, and the relationship between the nature park and forest management is also controversial (see below). Further on, the local residents were also asked about how important they consider certain tasks of the national park. They could evaluate the different national park functions on a 1–5 Likert scale. The answers show that the protection of biological and geological values as well as the protection of the landscape are the most important tasks according to the local residents (Fig. 4.7). The result that the geological protection received such a high score can be traced back to the fact that these are karst areas, where the most important pillars of the local natural values are the elements of the geological heritage (caves, sinkholes, karren, gorges). The role played by the national park in tourism is also considered important by local residents in most places. On the other hand, the function of the national park in supporting scientific research is considered less important by the residents in each of the examined parks, and the educational function is also rated a little lower than the nature protection functions.

In most places, tourists are satisfied with the opportunities offered by the national parks. Most of them support developments that are considered environmentally friendly (new panorama points, new educational paths), and the majority of current visitors oppose environmentally less friendly developments (such as adventure parks, swimming pools, and entertainment facilities). Regarding the motivation of visitors, it is worth noting that in the examined parks the "national park" label (or, in the case of Apuseni NP, the "nature park" label) plays an important role in the destination selection of tourists (Fig. 4.8). The Aggtelek NP is an exception to this, as the caves of this area were known "for their own right" long before it was declared a national park, and people come here mostly to visit these caves regardless of whether it is a national park or not. However, in the other cases, roughly half of the visitors indicated that visiting a national (natural) park is a very important aspect for them when choosing a travel destination.

Based on the questionnaires and interviews, the conflicts that exist between the individual actors can be also outlined. The main problem in the relationship between the state and the park managers is the decreasing budget, which leads to the fact that there are often not enough jobs for

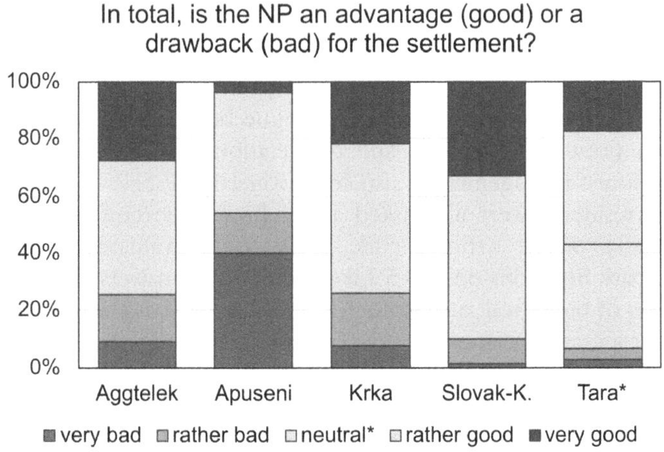

Fig. 4.6 Survey answers of local residents to the question "In total, is the NP an advantage (good) or a drawback (bad) for the settlement?" (*the "neutral" option was available only in case of the Tara NP survey)

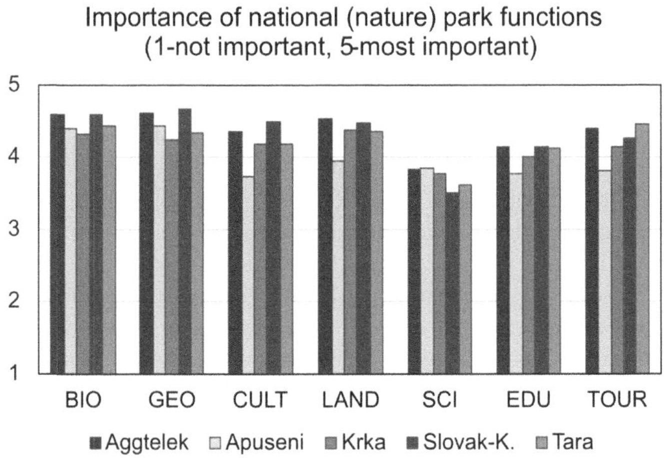

Fig. 4.7 Mean scores given to the importance of each task of the park according to local people. 1: not important at all, 5: very important. BIO: preservation of biological values; GEO: preservation of geological values; CULT: preservation of cultural values; LAND: landscape protection; SCI: scientific research; EDU: education; TOUR: tourism

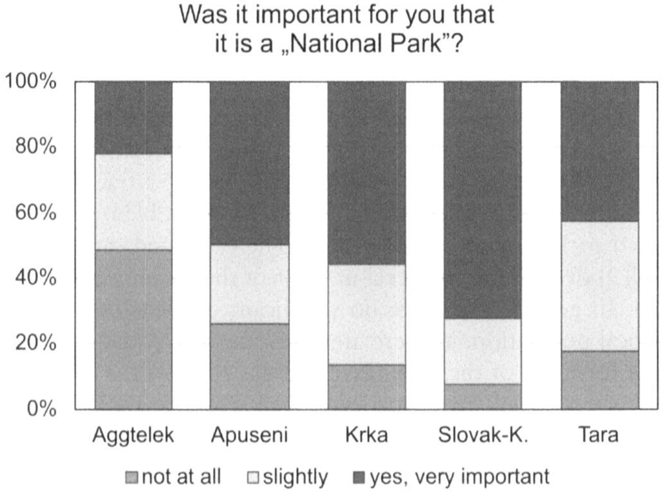

Fig. 4.8 Survey answers of tourists to the question “Was it important for you that it is a National Park?”

the park to fulfil its required tasks. The conflict between the park and the forestry is the management of the forests, during which nature conservation and economic aspects should be harmonized. In this regard, the situation of Tara NP seems to be conflict-free, where the forestry is actually a part of the national park. However, a sort of conflict exists between the NP management and local residents, because the locals are those who would like to get wood from the forests of the protected area. The Apuseni NP seems to be the most burdened, where the opposite is true, the nature park is subordinated to the state forestry, and local people feel that deforestation is too intensive and uncontrolled. In the relationship between the park and the local population, the following conflicts exist in most places: the park managers are mostly not from the settlements of the national park, thus they are considered “outsiders”. Further on, the restrictions related to nature conservation are often resented by local residents. Finally, in some places, disputes arise about the distribution of the benefits from tourism between the park and the local residents. As for the relationship between the park and the visitors, the issue of seasonality is a problem. In the summer period, there are crowds and congestion, while in the rest of the year there is too much “silence” (but this varies greatly from park to park). Solving seasonality in the studied parks seems to be a quite difficult problem. In addition, tourism is spatially highly inhomogeneous, as tourists usually concentrate on one or two prominent sites. In this issue, most of the examined parks try to do something in order to make the spatial distribution of visitors more uniform. However, the location and number of the most spectacular natural attractions are difficult to modify. Garbage is currently not a serious problem in most of the investigated parks (except Apuseni NP). The estimated carrying capacity is also well above the current level in most of the examined parks (except Krka NP). In general, there are no significant conflicts between tourists and the local population, as there are no disturbing crowds in the examined parks for most of the year (except Krka NP), so the local residents are rather happy about the tourists who contribute to their livelihood.

Finally, in relation to the attitude towards "karst" and "geotourism", it is concluded that for local residents, "nature" primarily means the surrounding forests, which they often visit. Caves are usually the biggest attraction for tourists in karstic parks (except Krka NP and Tara NP), however, for local people, they are less important for their daily life (except NP employees who are connected to caves by their work). In some places, but not everywhere, the national parks provide special programmes for the surrounding schools, through which the children living there can learn about the goals, tools, and important locations of the national park, or nature conservation in general. Local residents have a certain "field knowledge" about karst processes, as it affects their everyday lives. However, it is emphasized that the number of people who come into daily contact with the land through agriculture is gradually decreasing. According to the evidence of the surveys, most of the local residents have not yet heard of "geotourism". As for the tourists, their knowledge of karstification is generally not very deep. In most of the investigated places, less than half of the visitors claimed to know the meaning of the word "karst", and only about a quarter could correctly define the meaning of this word or list some typical karst landforms. At the same time, according to their own admission, more than half of the visitors knew the word "geotourism" in most places, and an average of 20–25% of the respondents declared themselves to be geotourists "to some extent".

Conclusions

The notion of "national park" and the order of goals has changed several times during the 150-year-long career of national parks. In Europe, the number of national parks increased continuously with an increasing rate since the beginning of the twentieth century until 2010. Most karst areas (in Europe) have always had relatively low population density, and depopulation is very typical nowadays. In general, the main goal of a national park is to protect nature, but it can also be a driving force in tourism—however, the benefits are not always enjoyed by local people. In the studied national parks (except Tara NP), geological and biological preservation tasks are considered equally important. For the local people, the national park is generally important as an economic potential (direct NP jobs; jobs in tourism), not because "nature" is so important in their lives. In the studied national parks, the majority of local residents have a "supportive attitude", but the proportion of opponents is also not negligible in some cases. Regarding the above conclusions, it is important to be aware that they are based on investigations carried out in national (natural) parks located in Eastern Central and Southern Europe. The examined countries are spatially close to each other, their current nature protection system and their historical past are also relatively similar (Mari et al. 2022). An important common element of their history is that until 35 years ago they belonged to the so-called socialist countries, where the role of the state was decisive and entrepreneurial freedom was strongly limited (Kőszegi et al. 2022). The impact of this past can still be recognized in the fact that people still see the "state power" in the institution of national parks. Nevertheless, the results with some modifications can be considered valid beyond the examined group of countries, presumably for the majority of karst national parks in Europe.

Acknowledgements This research was supported by the National Research, Development and Innovation Office Hungary (NKFIH) K124497 project.

References

Brankov, Jovana, Jasna Micić, Jelena Ćalić, Jelena Kovačević-Majkić, Ranko Milanović, and Tamás Telbisz. 2022. Stakeholders' attitudes toward protected areas: The case of Tara National Park (Serbia). *Land* 11: 468. https://doi.org/10.3390/land11040468.

Butler, Richard W., and Stephen W. Boyd, eds. 2000. *Tourism and national parks: Issues and implications*. Chichester: Wiley.

Clements, Reuben, Navjot S. Sodhi, Menno Schilthuizen, and Peter K. L. Ng. 2006. Limestone karsts of southeast Asia: Imperiled arks of biodiversity. *BioScience* 56: 733–742. https://doi.org/10.1641/0006-3568(2006)56[733:LKOSAI]2.0.CO;2.

Dowling, Ross K. 2011. Geotourism's global growth. *Geoheritage* 3: 1–13. https://doi.org/10.1007/s12371-010-0024-7.

Farsani, Neda Torabi, Celeste Coelho, and Carlos Costa. 2011. Geotourism and geoparks as novel strategies for socio-economic development in rural areas. *International Journal of Tourism Research* 13: 68–81. https://doi.org/10.1002/jtr.800.

Frost, Warwick, and Colin Michael Hall. 2015. *Tourism and national parks: International perspectives on development, histories and change*. London: Routledge.

Goldscheider, Nico, Zhao Chen, Augusto S. Auler, Michel Bakalowicz, Stefan Broda, David Drew, Jens Hartmann, et al. 2020. Global distribution of carbonate rocks and karst water resources. *Hydrogeology Journal* 28: 1661–1677. https://doi.org/10.1007/s10040-020-02139-5.

Gorjanc, Sa.šo, Tina Simončič, Aleš Poljanec, Béla. Kuslits, Ildikó Arany, Eszter Tanács, Ágnes. Vári, et al. 2022. A new ecosystem services approach to enable identification of pro-biodiversity businesses of protected karst areas in Central and South-Eastern Europe. *Hungarian Geographical Bulletin* 71: 181–195. https://doi.org/10.15201/hungeobull.71.2.6.

Grau, H. Ricardo., and T. Mitchell Aide. 2007. Are rural–urban migration and sustainable development compatible in mountain systems? *Mountain Research and Development* 27: 119–124. https://doi.org/10.1659/mrd.0906.

Gretter, Alessandro, Ingrid Machold, Andrea Membretti, and Thomas Dax. 2017. Pathways of immigration in the alps and carpathians: Social innovation and the creation of a welcoming culture. *Mountain Research and Development* 37: 396–405. https://doi.org/10.1659/MRD-JOURNAL-D-17-00031.1.

Imecs, Zoltán, András Máthé, and Balázs Kohán. 2022. Attitudes of local people towards Apuseni Nature Park, Romania. *Hungarian Geographical Bulletin* 71: 133–148. https://doi.org/10.15201/hungeobull.71.2.3.

Jamal, Tazim, and Amanda Stronza. 2009. Collaboration theory and tourism practice in protected areas: Stakeholders, structuring and sustainability. *Journal of Sustainable Tourism* 17: 169–189. https://doi.org/10.1080/09669580802495741.

Kohler, Thomas, Nodar Elizbarashvili, Giorgi Meladze, Davit Svanadze, and Heino Meessen. 2017. The demogeographic crisis in Racha, Georgia: Depopulation in the central caucasus mountains. *Mountain Research and Development* 37: 415–425. https://doi.org/10.1659/MRD-JOURNAL-D-17-00064.1.

Kőszegi, Margit, Alena Gessert, Janetta Nestorová-Dická, Péter. Gruber, and Zsolt Bottlik. 2022. Social assessment of national parks through the example of the Aggtelek National Park. *Hungarian Geographical Bulletin* 71: 149–162. https://doi.org/10.15201/hungeobull.71.2.4.

Kovačević-Majkić, Jelena, Jelena Ćalić, Jasna Micić, Jovana Brankov, Ranko Milanović, and Tamás Telbisz. 2022. Public knowledge on karst and protected areas: A case study of Tara National Park, Serbia. *Hungarian Geographical Bulletin* 71: 163–179. https://doi.org/10.15201/hungeobull.71.2.5.

Krpina, Vesna. 2015. Analysis of the relation between visitors and protected natural areas in the Zadar County. *Šumarski list* 139. Hrvatsko šumarsko društvo: 535–551.

Leroux, Shawn J., Meg A. Krawchuk, Fiona Schmiegelow, Steven G. Cumming, Kim Lisgo, Lee G. Anderson, and Mirela Petkova. 2010. Global protected areas and IUCN designations: Do the categories match the conditions? *Biological Conservation* 143: 609–616. https://doi.org/10.1016/j.biocon.2009.11.018.

Mari, László., Zsófia. Tábori, Ivan Šulc, Petra Radeljak Kaufmann, Ranko Milanović, Alena Gessert, Zoltán Imecs, Anetta Baricz, and Tamás Telbisz. 2022. The system and spatial distribution of protected areas in Hungary, Slovakia, Romania, Serbia and Croatia. *Hungarian Geographical Bulletin* 71: 99–115. https://doi.org/10.15201/hungeobull.71.2.1.

Mercier, Guy. 2009. Vidal de la Blache, P. In *International Encyclopedia of Human Geography*, ed. Rob Kitchin and Nigel Thrift, 147–150. Oxford: Elsevier. https://doi.org/10.1016/B978-008044910-4.00605-2

Mika, Mirosław, Bernadetta Zawilińska, and Magdalena Kubal-Czerwińska. 2019. Exploring the determinants of local people's attitude towards national parks in Poland. *Folia Geographica* 61: 5–16.

Milošević, Marko V., Milovan Milivojević, and Jelena Ćalić. 2010. Spontaneously abandoned settlements in Serbia, Part 1. *Journal of the Geographical Institute "Jovan Cvijić" SASA* 60: 39–57. https://doi.org/10.2298/IJGI1002039M.

Milošević, Marko V., Milovan Milivojević, and Jelena Ćalić. 2011. Spontaneously abandoned settlements in Serbia, Part 2. *Journal of the Geographical Institute "Jovan Cvijić" SASA* 61: 25–35. https://doi.org/10.2298/IJGI1102025M.

Mose, Ingo. 2007. *Protected areas and regional development in Europe: Towards a new model for the 21st century*. Aldershot, UK: Ashgate Publishing Ltd.

Dická, Nestorová, Alena Gessert Janetta, Lenka Bryndzová, and Tamás Telbisz. 2020. Behavioural survey of local inhabitants' views and attitudes about Slovak karst national park in Slovakia. *Sustainability* 12: 10029. https://doi.org/10.3390/su122310029.

Newsham, Niall, and Francisco Rowe. 2023. Understanding trajectories of population decline across rural and urban Europe: A sequence analysis. *Population, Space and Place* 29: e2630. https://doi.org/10.1002/psp.2630.

Papageorgiou, Kostas, and Kostas Kassioumis. 2005. The national park policy context in Greece: Park users' perspectives of issues in park administration. *Journal for Nature Conservation* 13: 231–246. https://doi.org/10.1016/j.jnc.2004.11.001.

Pejnović, Dane, and Dragica Husanović-Pejnović. 2008. Causes and consequences of demographic development in the territory of Velebit Nature Park, 1857–2001. *Periodicum Biologorum* 110: 195–204.

Pullin, Andrew S., Mukdarut Bangpan, Sarah Dalrymple, Kelly Dickson, Neal R. Haddaway, John R. Healey, Hanan Hauari, et al. 2013. Human well-being impacts of terrestrial protected areas. *Environmental Evidence* 2: 19. https://doi.org/10.1186/2047-2382-2-19.

Stolton, Sue, and Nigel Dudley. 2010. *Arguments for protected areas: Multiple benefits for conservation and use*. Oxford: Routledge.

Šulc, Ivan, and Valentina Valjak. 2012. Zaštićena područja u funkciji održivog razvoja hrvatskog otočja–primjer otoka Mljeta. *Croatian Geographical Bulletin* 74: 161–185.

Telbisz, Tamás, Christos L. Stergiou, Andrea Mindszenty, and Alexandros Chatzipetros. 2019. Geological and geomorphological characteristics of Vikos Gorge and Tymphi Mountain (Northern Pindos National Park, Greece) and karst-related social processes of the Region. *Acta Carsologica* 48: 29–42. https://doi.org/10.3986/ac.v48i1.6806.

Telbisz, Tamás, Ivan Šulc, László. Mari, and Petra Radeljak Kaufmann. 2022c. Attitudes and preferences of visitors of Krka National Park, Croatia. *Hungarian Geographical Bulletin* 71: 117–132. https://doi.org/10.15201/hungeobull.71.2.2.

Telbisz, Tamás, Jelena Ćalić, Jelena Kovačević-Majkić, Ranko Milanović, Jovana Brankov, and Jasna Micić. 2021. Karst Geoheritage of tara national park (Serbia) and its geotouristic potential. *Geoheritage* 13: 88. https://doi.org/10.1007/s12371-021-00612-5.

Telbisz, Tamás, Jovana Brankov, and Jelena Ćalić. 2020a. Topographic and lithologic controls behind mountain depopulation in Zlatibor District (Western Serbia). *Journal of Mountain Science* 17: 271–288. https://doi.org/10.1007/s11629-019-5861-5.

Telbisz, Tamás, and László. Mari. 2020. The significance of karst areas in European national parks and geoparks. *Open Geosciences* 12: 117–132. https://doi.org/10.1515/geo-2020-0008.

Telbisz, Tamás, László. Mari, Alena Gessert, Janetta Nestorová Dická, and Péter. Gruber. 2022a. Attitudes and perceptions of local residents and tourists—A comparative study of the twin national parks of Aggtelek (Hungary) and Slovak Karst (Slovakia). *Acta Carsologica* 51: 93–109. https://doi.org/10.3986/ac.v51i1.10683.

Telbisz, Tamás, Péter. Gruber, László. Mari, Margit Kőszegi, Zsolt Bottlik, and Tibor Standovár. 2020b. Geological heritage, geotourism and local development in Aggtelek national park (NE Hungary). *Geoheritage* 12: 5. https://doi.org/10.1007/s12371-020-00438-7.

Telbisz, Tamás, Petra Radeljak Kaufmann, and Neven Bočić. 2022b. Inland-coastal demographic transformations in a karst area: A case study of the surroundings of Krka National Park (Croatia). *Journal of Mountain Science* 19: 305–321. https://doi.org/10.1007/s11629-021-7032-8.

Telbisz, Tamás, Zoltán Imecs, László. Mari, and Zsolt Bottlik. 2016. Changing human-environment interactions in medium mountains: The Apuseni Mts (Romania) as a case study. *Journal of Mountain Science* 13: 1675–1687. https://doi.org/10.1007/S11629-015-3653-0.

Telbisz, Tamás, Zsolt Bottlik, László. Mari, and Alena Petrvalská. 2015. Exploring relationships between Karst terrains and social features by the example of Gömör-Torna Karst (Hungary-Slovakia). *Acta Carsologica* 44: 121–137. https://doi.org/10.3986/ac.v44i1.1739.

Telbisz, Tamás, Zsolt Bottlik, László. Mari, and Margit Kőszegi. 2014. The impact of topography on social factors, a case study of Montenegro. *Journal of Mountain Science* 11: 131–141. https://doi.org/10.1007/s11629-012-2623-z.

Trakolis, Dimitrios. 2001. Perceptions, preferences, and reactions of local inhabitants in Vikos-Aoos National Park, Greece. *Environmental Management* 28: 665–676. https://doi.org/10.1007/s002670010251.

Watson, James E. M., Nigel Dudley, Daniel B. Segan, and Marc Hockings. 2014. The performance and potential of protected areas. *Nature* 515: 67–73. https://doi.org/10.1038/nature13947.

Zachrisson, Anna, Klas Sandell, Peter Fredman, and Katarina Eckerberg. 2006. Tourism and protected areas: Motives, actors and processes. *International Journal of Biodiversity Science and Management* 2: 350–358. https://doi.org/10.1080/17451590609618156.

Zawilińska, Bernadetta. 2020. Residents' attitudes towards a national park under conditions of suburbanisation and tourism pressure: A case study of Ojców national park (Poland). *European Countryside* 12: 119–137. https://doi.org/10.2478/euco-2020-0007.

Zgłobicki, Wojciech, and B. Baran-Zgłobicka. 2013. Geomorphological heritage as a tourist attraction. A case study in Lubelskie Province, SE Poland. *Geoheritage* 5: 137–149. https://doi.org/10.1007/s12371-013-0076-6.

Zurc, Jana, and Andrej Udovč. 2009. Local inhabitants' opinion about the Triglav national park management. *Sociologija i Prostor/sociology and Space* 47: 43–56.

CHAPTER 5

National Parks and Wildlife Restoration in the Global North: The Case of Wolves in the United States and France

Randall K. Wilson and Alyssa Papantonakis

Introduction

National parks have played a central role in efforts to protect and restore endangered and other at-risk wildlife species in the United States. As the hub of the Greater Yellowstone Ecosystem, Yellowstone National Park has rendered a profound impact on some of the earliest attempts to protect bison, elk and many other species, including gray wolves (*Canis lupus*) (Wilson 2020).

R. K. Wilson (✉)
Department of Environmental Studies, Gettysburg College, Gettysburg, PA, USA
e-mail: rwilson@gettysburg.edu

A. Papantonakis
Department of Urban Studies and Planning, Massachusetts Institute of Technology, Cambridge, MA, USA
e-mail: alyssap7@mit.edu

J. Weber and S. Sultana (eds.), *The Changing Geography of National Parks and Protected Areas*,
https://doi.org/10.1007/978-3-031-74653-6_5

Through much of the long history of western civilization, wolves have been viewed in a negative light, if not outright demonized (Fogleman 1989). Their ability to prey on a wide array of species, including livestock, means wolves have typically been treated as a threat to economic prosperity as well as to human safety. However, as keystone species, they also serve as a vital component of healthy ecosystems. Situated at the very top of the food web, their predation patterns can impact each trophic level beneath them in a cascading effect (Smith et al. 2020). Ultimately, the loss of top predators can result in an overall loss of biodiversity. The near eradication of wolves, followed by their reintroduction has been a key part of Yellowstone National Park's legacy in wildlife protection and restoration in the United States.

This historical progression—from species extermination to eventual restoration and proliferation—reflects the history of wolves in many countries around the world. It also reflects the fact that wolves are one of the most widely distributed mammalian species on earth, ranging from northern Canada to southern Mexico, and from southern India to northern Russia and even Greenland (Young and Goldman 1964; Mech and Boitani 2003).

The history of wolves in France is no exception. The timeline is remarkably similar to that in the United States, comprising a period of near eradication during the early twentieth century followed by a return in the early 1990s (Moriceau and van Maanen 2023). Temporal parallels between the US and French experience, along with socio-economic and geographic similarities across the two primary regions in which reintroduction has taken place, offer an opportunity for comparative analysis. Despite many similarities, the United States' and France's respective governments have taken somewhat divergent approaches to wolf management, leading to differences in terms of policies and the resulting social, economic and ecological outcomes. Each country has unique perspectives to offer to this issue as they both work to better understand effective wildlife management policies. Our interest lies not only in wolf restoration, but in the role of national parks and protected areas as part of these efforts.

National parks and protected areas around the world are often viewed as being modeled on Yellowstone, but to what extent has Yellowstone offered a model for wildlife conservation? Is Yellowstone's legacy a function of its unique history and geography? Or does it offer a potential template for wildlife conservation and restoration in other parks around the world? Conversely, what lessons for park and wildlife management do other countries hold for the United States? This chapter explores these questions through a preliminary investigation of national parks and wolf restoration in the United States and France.

Adopting an comparative case study approach (Yin 2017) we examine (1) the historical trajectory of policy and management with regards to wolves in the two countries, (2) the geographical context, both physical and human, in which restoration is taking place, (3) the ecological implications of restoration efforts for wolves and other wildlife, and finally, (4) socio-economic implications in terms of livestock depredation, ecotourism and sport hunting. Data is drawn from both primary and secondary sources, including wildlife and agricultural census data, and government issued policy documents.

Historical Trajectories

Through much of US history, wolf management really meant eradication. As early as 1630, colonial authorities in the Massachusetts Bay Colony offered a one shilling bounty as a cash reward for each wolf killed (Foster et al. 2002). This set an early precedent of discouraging the presence of wolves and set the stage for a series of anti-wolf management policies. When Yellowstone National Park was first established in 1872 to protect wildlife from "wanton destruction," protections did not extend to predator species. Early park managers viewed wolves as a threat to the game species they wished to preserve, including elk, pronghorn, deer and some of the last surviving herds of plains bison. Ultimately, predator control programs supported by the Park Service evolved into policies of systematic extermination. By the time the last wolf pack in Yellowstone was killed in 1926, the only surviving gray wolf packs could be found in northern Minnesota and what later became known as Isle Royale National Park in Michigan (Smith et al. 2020). Within Yellowstone, the absence of wolves rendered significant impacts on flora and fauna throughout the park, requiring mangers to engage in heavy culling in order to maintain ungulate populations below perceived carrying capacities within park boundaries.

Though wolves no longer lived within the boundaries of Yellowstone National Park, in 1978, the federal government listed the gray wolf as endangered in nearly all states under the Endangered Species Act. In the late 1980s, a pack of wolves arrived from Canada via natural migration into Glacier National Park in Montana. Though later eradicated, these wolves forced a public debate on restoration. In 1994, after 20 years of political struggle, a coalition of wolf restoration advocates, including the National Park Service, succeeded in initiating a process to restore gray wolves to the northern Rockies. Ironically, this approach was supported in part because under the ESA's "non-essential experimental population" categorization, human-induced reintroduction allowed residents greater opportunity to protect livestock against wolf predation than would be possible under restoration caused by natural migration. In 1994 and 1995, NPS and USFWS officials relocated thirty-one wolves from western Canada into Yellowstone National Park and to a wilderness area in Idaho. These two populations, in conjunction with a naturally returned population in northern Montana, comprised a multi-pronged conservation approach centered on the Greater Yellowstone Ecosystem.

Ecologically speaking, the return of wolf populations to Yellowstone National Park has been largely successful. Nonetheless, since 2008, the wolves of the Northern Rocky Mountains Distinct Population Segment have been delisted and relisted numerous times under the Endangered Species Act, reflecting the persistent controversy surrounding this species. Diverse political and economic concerns have been reflected in state-level management plans, as each state has taken a different approach to the issue. Some states have adopted protective measures for the wolves in coordination with Yellowstone's goals while others have deviated at times from national park priorities. Regardless, the wolf population originally established in Yellowstone has continued to expand beyond park boundaries (Mech 2017). Today, the wolves within the park are fully protected, and those of the Greater Yellowstone Ecosystem are managed through the state management plans of Wyoming, Montana and Idaho, all of which allow regulated hunting.

The French story is remarkably similar to that of the United States. However, the history is much longer, contributing to the development of European cultural values that would lead to early anti-wolf policies in both countries. In approximately 813 AD, Charlemagne established the *luparii*, a group of trained wolf hunters designed to protect French livestock and citizens (Bernard 1981; van Maanen 2023). Later evolving into the *Louvertier Royal*, this institutionalized extirpation effort along with a bounty system existed through much of the *Ancien Regime* and continues to this day (albeit in modernized form). Medieval stories such as "Les Loups de Paris" and the "Beast of Gévaudan," describing alleged attacks and terrorization of the French people, added to wide-spread beliefs that wolves posed a threat to rural livelihoods (Thompson 1992). Despite few documented wolf attacks on humans, the psychological impact of these offered further support for the extirpation of wolves. Although France had been home to one of the largest wild wolf populations in Europe prior to the nineteenth century, numbering somewhere between 10,000 and 15,000 wolves, the population dropped to near zero by the early 1930s (Moriceau 2013).

In 1963, France established its first national park: Vanoise National Park. As a response to the near loss of ibex in the Alps, the creation of Vanoise represented a point of progress for French wildlife conservation (Maus 2007). This spurred the creation of additional national parks in France to protect wildlife and conserve natural landscapes, including the establishment of Mercantour in 1979.

That same year, the Council of Europe initiated the first protections for wolves and other large carnivores in Europe. Opened for signatures in 1979 and coming to force in 1982, the Bern Convention on the Conservation of Wildlife and Natural Habitats in Europe called for the conservation of wild flora and fauna and gave special protection to endangered species. France ratified the treaty in 1990, which outlawed the killing of wolves unless they posed a direct threat to humans or livestock. In 1992, the European Union Habitats Directive reaffirmed this decision, protecting gray wolves as species included on IUCN Red List. As there were no known wolves present at this time in France, it was not difficult for French authorities and residents to both support and comply with these directives. However, the situation would soon change.

Just like in the United States, wolves first returned to France via natural migration. They first appeared in Mercantour National Park in 1993, arriving from the Abruzzo Mountains in Italy (Poulle et al. 1999). Hence, they did not represent a purposeful reintroduction effort as seen in Yellowstone and Idaho. Nonetheless, both Mercantour and Vanoise National Parks have served as important hubs for wolf restoration. Located along the French Alps, both parks have offered wolves safe havens from hunting, thereby allowing the newly arrived wolf populations to grow and expand into other parts of France (Boitani et al. 2018).

Consequently, and again, similar to the US experience, as wolves have continued to migrate outside of park borders in France, conflicts with local residents and livestock producers have grown. Beginning in 2008, the French government implemented its first National Action Plan (NAP) to address growing conflicts and develop national goals and standards for wolf management. Led by French national ministries associated with ecological conservation and agriculture, governmental authorities also relied on public input and consultation from the National Wolf Group, an informal organization that brought together stakeholders on various sides of the issue to mediate conversations, conduct assessments and contribute to the development of future plans. Through these national plans, updated every 5 years, the government aims to balance the protection of wolves and livestock by managing for a specific wolf population size. According to France's initial NAP, wolves could be killed in "extreme cases"—meaning they posed a direct threat to humans or livestock. However, in 2012, the French government adopted a new approach designed to limit population growth. The 2018–2023 plan sought to maintain a national population of approximately 500 wolves through a mix of hunting, defensive kills by farmers, and exterminations ordered by the *louveterie* (Benhammou and Degeorges 2023). The official justification for this policy shift was the rapid growth in the number of human-wolf conflicts (primarily expressed as livestock predation). Consequently, in recent years the French government has further lessened its protective measures for the wolves in order to safeguard agricultural interests.

The proposed 2024–2029 NAP, however, angered both conservation organizations and farmers (FERUS 2022). Few if any of the suggestions made by environmental groups were reflected in the final version of the plan, with the government instead placing an emphasis on management through culling. In response, six environmental associations withdrew from the National Wolf Group (News Wires 2023). However, the agricultural community also remains frustrated, concerned by the continued threat wolf population poses to livestock as exemplified through increasing livestock kills. The new plan proposes an increased culling rate of up to 19 percent of the current estimated population, enabling the extermination of up to 209 wolves in 2024 (Benhammou and Degeorges 2023).

Physical Geography

For the purposes of this study, rather than compare the entire geographic area of the United States with that of France, when the data allow, we focused primarily on the area comprising the Greater Yellowstone

Ecosystem (GYE) in the United States and the regions of Provence-Alpes-Côte-d'Azur and Auvergne-Rhône-Alpes in France (Fig. 5.1). Together, these regions present geographies of relatively similar size which also serve as the areas experiencing the most intensive wolf restoration in each country.

The focal points for wolf reintroduction in the United States included Yellowstone National Park and the Frank Church River of No Return Wilderness in Idaho (as well as naturally returning packs in northern Montana). The park encompasses 8991 square kilometers of land, ranging from 1610 meters to 3462 meters in altitude. It is a central part of the GYE, which extends approximately 89,000 square kilometers and includes lands in the states of Idaho, Montana and Wyoming (Keiter 2020).

The Greater Yellowstone Ecosystem includes several protected areas ideal for wildlife conservation, including two national parks (Yellowstone and Grand Teton), five national forests (Bridger-Teton, Shoshone, Caribou-Targhee, Gallatin-Custer, and Beaverhead-Deerlodge) and the National Elk Refuge. Each of the large iconic big game and predatory species present in the early 1800s persist in the ecosystem today, but wildlife management practices have been a point of contention throughout the region's history (Keiter 2020). Abundant populations of

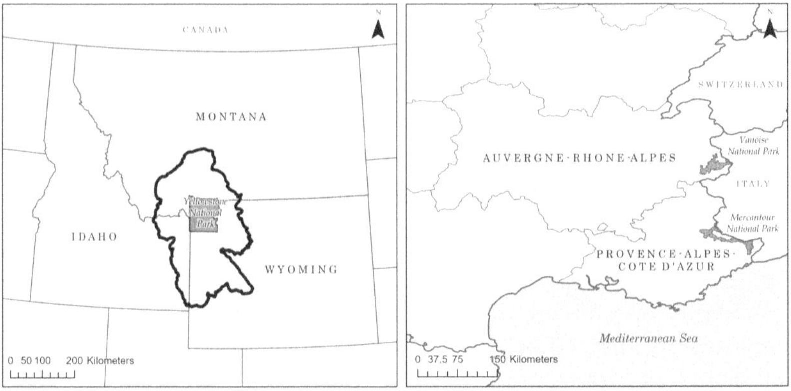

Fig. 5.1 Map of the greater Yellowstone ecosystem (United States), showing Yellowstone national park and the Auvergne-Rhône-Alpes and Provence-Alpes-Côte D'Azur region in France, showing Vanoise and Mercantour National Parks. *Source* Map by the authors

elk provide a particularly important source of natural prey for the wolves in this region.

The country of France has a total area of 551,696 square kilometers, roughly aligning with the combined area of the states of Wyoming and Montana (634,432 sq km). Provence-Alpes-Côte-d'Azur and Auvergne-Rhône-Alpes are 31,400 square kilometers and 69,711 square kilometers respectively. These two regions border each other, spanning a combined area similar to that of the Greater Yellowstone Ecosystem at roughly 101,111 square kilometers. Together, the regions make up the south-eastern portion of the country alongside the Italian border. This area represents a wide variety of landscapes, marked by mountains to the east, the Mediterranean Sea to the south, and rolling valleys throughout. Important prey within the Auvergne-Rhône-Alpes and Provence-Alpes-Cote-d'Azur regions include wild ungulates such as deer, mouflon, ibex and young boar. Mouflons are not native to the region and were instead introduced to mainland Europe from the islands of Corsica and Sardinia for hunting purposes (Tizzani et al. 2022). Due to the mouflon's poor adaptation to snow and artificiality to the region, this species has become an easy source of prey for wolves (Poulle et al. 1999; FERUS 2022).

Protected areas in the two French regions constitute a much smaller portion of the landscape than in the Greater Yellowstone. Vanoise National Park, located within the Auvergne-Rhône-Alpes region, has served a similar role to Yellowstone in terms of bringing national attention to wildlife protection. The park consists of only 528 square kilometers, but its borders connect with the Italian Gran Paradiso National Park to produce a total protected area of 1250 square kilometers. Mercantour National Park, located within the Provence-Alpes-Côte-d'Azur, was the site of the first confirmed wolf sighting after their return in the 1990s. This biodiverse landscape spans seven valleys and numerous peaks, reaching a maximum altitude of 3143 meters and stretching across approximately 685 square kilometers in area.

In terms of land use and human populations, the GYE is largely comprised of conservation lands and rural agricultural areas. The population density of Montana is 2.95 people per square kilometer. The figure for Wyoming is 2.29, and Idaho comes in at 8.95. Several of these states, however, are undergoing rapid population growth. Idaho experienced the second highest state-level population growth rate in the United States between 2010 and 2020 (17.3%) while Montana was number fourteen (9.6%) (US Census 2024). The primary population centers within the

GYE include Jackson, Pinedale, Cody and Afton in Wyoming; Driggs, Victor and Idaho Falls in Idaho; and West Yellowstone, Red Lodge and Bozeman in Montana (Keiter 2020).

The states of Montana, Wyoming and Idaho each contain a significant portion of agricultural land, totaling 39.8 million hectares between the three states. All three states dedicate much of their agricultural land to cattle. These farms are typically large-scale, commercial farms, averaging 875.7 hectares per farm in Montana, 1103.2 hectares per farm in Idaho and 189.2 hectares per farm in Idaho (USDA 2024), all of which offer wolves the potential for prey, habitat and the possibility of conflict with humans.

In France, wolves encounter a much more densely populated landscape than in the GYE. As of 2022, France has a total population of 65.8 million people, with a population density of 119.3 persons/sq km across the country. The regions of Provence-Alpes-Côte-d'Azur and Auvergne-Rhône-Alpes are even more densely populated, at 163.4 people/sq km and 259.6 people/sq km respectively (OECD 2022). While this area includes rural towns and villages like those scattered throughout the GYE, it also contains much larger metro areas including Nice-Cannes, Marseilles-Aix-en-Provence, and Lyon. Removing the populations and geographical area of these major metro areas, as wolves generally do not occupy these spaces, the total population of this region is approximately 8,971,000, producing a population density of approximately 99.2 persons/sq km in the region (OECD 2022). Insofar as the wolf populations tend to occupy rural areas rather than urban ones, this data serves to qualify the density measures somewhat and lessen the distinctions between the United States and France (Mech 2017). Still, the population density in this region of France remains much greater than that of the Greater Yellowstone Ecosystem.

On the other hand, just over half (50.8%) of all land in France is farmland, representing approximately 456,000 farms. These farms are much smaller than those in the GYE, averaging 69 hectares in size. French farms produce a variety of agricultural products, including grains, root vegetables, and livestock. Auvergne-Rhône-Alpes is the French region with the greatest number of food-industry businesses as well as the second greatest area devoted to organic farming. As in the GYE, the prevalence of livestock in the region provides a potential food source for wolves, setting the stage for human conflict.

Outcomes and Impacts

While there are some notable differences between the United States and France in terms of the evolution of wolf restoration policies, many similarities remain. In both cases, one can argue, national parks played a prominent, if early, role in the development of this issue. And in both cases, the wolf populations have grown and expanded into geographic areas that are largely rural with relatively low human population densities (though densities are relatively higher in the French case). In what follows, we consider how the ecological, social and economic implications of restoration have played out, including variations in policy adaptations and responses.

Wolf Population Growth

According to its wolf recovery plan, Yellowstone National Park aims to maintain its annual wolf population at roughly 100 wolves. As noted in Table 5.1, in 2022, Yellowstone National Park had a total of 108 wolves, consisting of 10 packs and 7 breeding pairs (Smith et al. 2020). The Greater Yellowstone Ecosystem, including the area surrounding the park within the states of Idaho, Montana and Wyoming, has a wolf population of approximately 500 individuals. According to the ESA, the states of Wyoming, Montana and Idaho each must maintain or exceed the minimum recovery criteria to ensure gray wolves are not in danger of becoming relisted. As of 2022, Montana remained well above its minimum management goals with a population of 1087 wolves (Parks et al. 2023). Similarly, Idaho's wolf population in 2022 reached 1337, though the state aims to reduce this population to 500 in the coming years (Idaho Dept. of Fish and Game 2023). As for Wyoming, after the rejection of early plans that did not meet minimum federal standards, the state also met its conservation goals. In 2022, Wyoming hosted a population of 338 wolves with twenty-three breeding pairs (Mills 2023).

In France, the 2018 National Action Plan defined a target population size of 500 wolves by 2023, but the wolf population quickly surpassed this amount. By 2022, the total wolf population in France reached 1104 (Reed 2023). While most of these wolves are in southeastern France, the animals have been spotted in about half of France's 94 departments (Meuret and Regnier 2021). The total number of wolves in France is similar to the total count across the State of Montana.

Table 5.1 Wolf Populations in France and the Greater Yellowstone States. *Source* Idaho Dept. of Fish and Game, 2023; Mills, 2023; OFB, 2021; Parks et al., 2023)

Year	*France*		*Montana*		*Idaho*		*Wyoming*		*YNP*	
	Pop	*Density (per km^2)*	*Pop*	*Density (per km^2)*	*Pop*	*Density (per km^2)*	*Pop*	*Density (per km^2)*	*Pop*	*Density (per km^2)*
2015	280	0.00051	1,190	0.00312	786	0.00362	382	0.00151	98	0.01090
2016	360	0.00065	1,126	0.00295	n/a	n/a	377	0.00149	108	0.01201
2017	430	0.00078	1,117	0.00293	n/a	n/a	347	0.00137	97	0.01079
2018	530	0.00096	1,153	0.00303	1,543	0.00712	286	0.00113	80	0.00890
2019	580	0.00105	1,159	0.00304	1,270	0.00586	311	0.00123	94	0.01045
2020	624	0.00113	1,184	0.00311	1,556	0.00718	327	0.00129	123	0.01368
2021	921	0.00167	1,143	0.00300	1,543	0.00712	314	0.00124	97	0.01079
2022	1,104	0.00200	1,087	0.00285	1,337	0.00617	338	0.00133	108	0.01201

As far as the ecological impacts of wolf restoration is concerned, much has been written about the "trophic cascade" effect inside Yellowstone National Park. While still debated, early evidence suggests that the renewed presence of top predators has brought significant change to the Yellowstone ecosystem, which had suffered degradation from the overabundance of game species, most notably, elk (Smith et al. 2020). Wolf predation on elk has not only reduced elk numbers, but produced ripple effects across the ecosystem, allowing for the recovery of aspen, cottonwoods, willows and a variety of interrelated flora and fauna.

Outside park boundaries, there is less agreement on the ecological effects of wolves. Most agree that they have led to a reduction in the regional elk populations, but there is much debate as to their specific contribution given the many other variables involved (Smith et al. 2016). In France, the situation is also somewhat murky. There is some evidence that populations of non-native mouflon have dropped considerably since wolf restoration began (FERUS 2019). Like elk in the GYE, these species thrived for decades in Alpine habitats devoid of top predators. However, the drop in their population may be contributing to the rise of wolf predation on livestock.

Socio-Economic Issues

Livestock Depredation

Although historically, wolves prey primarily on wild ungulates, in places characterized by high livestock densities and a limited supply of wild prey, wolves will increasingly feed on domestic livestock. Though predators other than wolves along with additional factors, such as weather or disease, account for most livestock loses, wolves frequently bear a disproportionate share of the blame. For example, in 2020, wolves accounted for just 1.0 percent of sheep and lamb deaths in Montana. Far more sheep and lambs were killed by old age (30.0%), coyotes (16.0%), weather conditions (7.0%), birthing complications (7.0%) and diseases (6.0%) (Sommer 2021). Data in the state of Wyoming tells a similar story. In 2022, wolves caused just 0.5% of sheep and lamb deaths, while coyotes accounted for 32.4%, weather conditions (19.8%) and birthing complications (9.3%) (Lohrenz 2023). In the Greater Yellowstone Ecosystem, large populations of wild ungulates in conjunction with lesser densities of local livestock across a rather wide geographic area have resulted in relatively low depredation rates compared with those found in France (Table 5.2). Nonetheless, it is still a cause for concern among regional livestock growers, strong enough to have produced some innovative policy responses.

Although the "experimental population" categorization of wolves under the ESA allowed ranchers to protect their livestock from attacks by lethal means from the very beginning, the development of a compensation program to reimburse individual farmers for their losses emerged shortly thereafter. In 2010, the federal government took over the program, which

Table 5.2 Total Livestock Depredation, 2017 to 2022. *Source* DREAL, 2022, OFB, 2023, USDA, 2024

Year	*France*	*Provence-Alpes-Cote d'Azur*	*Auvergne-Rhone-Alpes*	*Montana*	*Idaho*	*Wyoming*
2017	11,936	7,586	3,320	80	184	194
2018	12,515	7,984	3,222	71	297	71
2019	12,451	7,747	3,734	94	313	70
2020	11,366	6,213	3,971	99	211	76
2021	10,348	5,732	3,011	96	198	109
2022	10,728	4,740	4,232	103	279	97

was later transferred to the states. In 2022, Montana provided livestock growers with $96,545.50 and Wyoming paid $187,382 in compensation. In addition, in 2022, 66 wolves were legally killed across the two states in response to livestock deaths (Parks et al. 2023; Mills 2023). The primary expenditure in wolf management then is a combination of compensation to farmers and the cost of state-enacted wolf extermination, actions taken when the state decides a problem wolf or wolf pack must be put down.

By contrast, France has experienced a much higher number of livestock losses caused by wolf predation. In 2022, the country recorded 10,853 total livestock killed, compared to only 479 in all of Montana, Wyoming and Idaho combined. Most of these attacks occurred in Provence-Alpes-Côte-d'Azur (4740) and Auvergne-Rhône-Alpes (4232). Again, this is an area comparable to the GYE and with a wolf population of comparable size. However, the total livestock population and density is much greater in the Provence-Alpes-Côte-d'Azur and Auvergne-Rhône-Alpes regions than across the states of Montana, Wyoming and Idaho. Additionally, while wolf populations have significantly increased in recent years, the total count of livestock kills remains relatively stagnant.

The vast majority of the French livestock deaths (91.5%) are associated with sheep. Even still, just 0.149 percent of the 6.6 million living sheep in France were killed by wolves in 2022 (Table 5.3). While the total counts of livestock kills are much greater in France than in the states of Montana, Wyoming and Idaho, the rates of depredation as compared to the total livestock populations are only slightly greater in France than in the Greater Yellowstone region. As in the United States, the percentage of livestock deaths caused by wolves pales in comparison to other causes of death. Across France, approximately 750,000 sheep die each year, mostly due to disease, parasites and accidents. Wolf predation accounts for approximately 1.3 percent of these deaths.

In response, France has utilized both preventative measures and compensatory efforts to livestock growers. Annually, the country spends approximately thirty-eight million euros on these efforts, with about 90% going to herd protection measures such as fencing, guard dogs, and supplementary salaries for herders and wolf extermination programs. The remaining 10 percent goes to farmers as compensation after a kill (Meuret and Regnier 2021; Ministère Écologique 2024). These pre-emptive measures to limit livestock deaths reflect the greater impact wolves have had on livestock producers in this region compared to the United States. The Auvergne-Rhône-Alpes and Provence-Cote-d'Azur

Table 5.3 Cases of Livestock Depredation (Sheep), 2017 to 2022. *Source* DREAL, 2022; OFB, 2023; Idaho Dept. of Fish and Game, 2023; Lohrenz, 2023; Mills, 2023; OFB, 2021; Parks et al., 2023; USDA, 2024

Year	*France*			*Montana*			*Wyoming*			*Idaho*		
	Killed by Wolves	*Total Population*	*Percent*	*Killed by Wolves*	*Total*	*Percent*	*Killed by Wolves*	*Total*	*Percent*	*Killed by Wolves*	*Total*	*Percent*
2017	11,312	6.877 mil	0.164%	12	230,000	0.005%	81	360,000	0.023%	93	250,000	0.037%
2018	11,789	7.166 mil	0.165%	7	225,000	0.003%	15	345,000	0.004%	169	230,000	0.073%
2019	10,374	7.105 mil	0.146%	21	215,000	0.009%	27	350,000	0.008%	123	220,000	0.056%
2020	10,348	6.998 mil	0.147%	32	200,000	0.016%	12	340,000	0.004%	79	230,000	0.034%
2021	9,494	6.995 mil	0.136%	29	200,000	0.015%	53	340,000	0.016%	81	230,000	0.035%
2022	9,816	6.598 mil	0.149%	41	190,000	0.022%	46	330,000	0.014%	192	N/A	N/A

regions not only reflect higher overall human population densities than can be found in the Greater Yellowstone Ecosystem, but also lower densities of wild prey and higher densities of small farms and domesticated animals.

Ecotourism
While the economic impacts of wolf reintroduction often focus on the negative effects of predation, the presence of wolves can also contribute positively to local economies through ecotourism. In the United States, early estimates predicted that wolf recovery in the GYE would create economic benefits of between $6.7 million and $9.9 million annually from the increase in tourism. Following reintroduction, wolves have had an even greater positive impact. Based on survey respondents stating that the possibility of seeing wolves positively impacted their decision to visit Yellowstone National Park, Duffield et al. (2008) estimated that the presence of wolves has generated approximately 35 million dollars annually in tourist revenue to the states of Montana, Idaho and Wyoming. Many visitors are more inclined to visit the park due to the possibility of observing wolves, and about 20% of visitors to the park purchase wolf-related items on their trips (Duffield et al. 2006). Estimates based on the assumption that a similar proportion of visitors are influenced by the presence of wolves predicted that the positive economic impact from ecotourism has continued to grow in recent years (ITRR 2022). However, there is a need to confirm these findings with updated and more targeted research.

The cascading impacts of wolves on the ecosystem may have impacted other forms of ecotourism in the GYE as well. A reduction of browsing by elk and deer in riparian zones may have increased vegetation and decreased stream temperatures. The resulting improvements in conditions for cold-water fish, such as cutthroat trout, may have benefitted the local sports fishing industry (Weiss et al. 2007).

Within France's Mercantour National Park lies Parc Alpha, a wildlife park and educational center established in 2005. Three packs of wolves (the Boréon pack, the Erps pack and the Pelago pack) roam within the park and can be observed by tourists. As with the deliberate relocation of wolves into the Greater Yellowstone Ecosystem, the wolves in Parc Alpha were brought to the park from other regions through exchange programs to promote and conserve genetic diversity. The wolves of the Boréon pack are initially from Canada, those of the Pelago pack come from the Arctic, and those of the Erps pack come from across Europe (Gauvin 2013).

This park is one of the most prominent tourist attractions along the Côte d'Azur, bringing 66,422 visitors in 2018 (CRT Côte D'Azur 2020).

The majority of the French general public (51.9%) agrees that wolves have a positive impact on tourism. This number increases among French tourism professionals to nearly two thirds (65.9%). However, there is limited clarity on this issue, with a significant number of survey respondents in both groups reporting a neutral reaction to this question (34.8% and 22.7% respectively) (Simon et al. 2023).

As in the United States, there is a need for additional research on the impacts of wolf populations on ecotourism. It is clear that wolves play a role in tourism to the national parks and to their surrounding areas, but data on specific economic impacts remains sparse.

Hunting

The issue of sports hunting cuts in several different policy directions. On one hand, it concerns hunting advocates worried that wolf reintroduction has reduced the number of elk or other desirable big game hunting species. This is one major rationale, especially in the United States, for opposing wolf restoration. On the other hand, wolf restoration offers the opportunity for private citizens to engage in the legal hunting of a new top predator.

In the United States, hunting is prohibited within the boundaries of Yellowstone National Park, but is allowed in national forests and other state and federal lands located within the GYE. If wolves from Yellowstone leave park boundaries, they may be hunted through these policies. In 2022, twenty-one wolves originating from the park were killed for that reason (French 2023).

In the Greater Yellowstone Ecosystem, hunting is managed on a state-by-state basis. In 2021, the state of Montana enacted legislation to increase wolf kills and reduce overall wolf populations through increased limits, longer trapping seasons, allowing the use of private funds to reimburse hunters, the use of snares to catch animals, the use of bait, and enabling hunters to kill wolves on private land at night (French 2023). However, while the state set its hunting and trapping quota at 450 in the 2023 season, just 258 wolves were killed that year. This represented a decline from the previous year's harvest of 273 wolves. As hunters are not always successful, these numbers demonstrate that private hunting is not necessarily an effective means for controlling or limiting wolf populations. Nonetheless, the sale of hunting licenses in Montana has a positive economic impact. Each year, from 2020 to 2022, Montana earned approximately $300,000 in state revenue (Parks et al. 2023).

In 2021, the state of Idaho similarly enabled an expansion in wolf hunting practices and permits. New laws removed limitations on the number of wolf tags hunters could obtain, enabled the use of tax revenue to hire private contractors to kill wolves, and expanded the length of the wolf hunting season (Maldonado 2023). Nonetheless, the state of Idaho saw a decrease in the number of harvested wolves from 438 in 2021 to 351 in 2022.

In Wyoming, mortality limits are set much lower due to a lower total wolf population. In 2022, the state of Wyoming sold 2323 wolf hunting licenses, resulting in 29 legal kills. Although there were also 2 illegal kills of wolves in the state, these numbers remained well below the set mortality limit of 47 (Mills 2023).

Assessing the effects of wolf mortality on livestock depredation across these three states of Idaho, Montana and Wyoming from 1987 to 2012, Wielgus and Peebles (2014) found that greater wolf hunting rates actually resulted in higher depredation rates the following year. The authors attributed this impact to the separation of packs, causing individual wolves to form new pairs and breed at higher rates. The study suggested that lethal wolf control, while potentially effective in the short-term, could be associated with an increase in future depredation.

In France, due to the 1979 Bern Convention and the 1992 European Habitats Directive, all wolf populations are legally treated as protected species, with only limited exemptions allowed for certain populations (European Commission 2023). Shortly thereafter, France moved away from a model of strict protection as French farmers began to experience significant losses due to livestock predation. In 2004, France's first national wolf plan allowed for a total of just six wolves to be legally killed each year and provided additional permits for extreme cases of predation when no other methods were feasible (France 24 2011). However, over the next 7 years, a grand total of just six wolves were legally hunted. At this point, only targeted kills were allowed on a case-by-case basis following notable wolf attacks. The policy changed dramatically with the 2012–2017 National Wolf Plan, which included a shift to manage for an overall wolf population size (Benhammou and Degeorges 2023). In the 2012–2017 plan, annual wolf culling rates were set at 10–12 percent of the total estimated population, under the belief that this aligned with annual population growth rates. But unanticipated increases in the wolf population meant the number of legal wolf kills grew each year, from 12 in 2012 to 40 in 2018 (Huisman 2018). In 2019, France allowed 96 kills, representing a 19% harvest rate (Huisman 2019), which continued through 2020. The latest national plan (for 2024–2029) adopts an even higher culling rate of 21 percent (Behammou and Degeorges 2023). Further, the plan relaxes constraints on methods allowed for killing wolves, enabling the use of night vision aids and hunting in pairs in addition to implementing an online platform that will simplify the deployment process for wolf hunters (Henshell 2023). The constant increase in culling rates reflects not only the robust growth of the French wolf population but the impact of public opinion on the issue of restoration.

As in the United States, some organizations have come to realize the limitations of hunting as a method to protect livestock populations. In many cases, the results of culling provide relief from depredation for certain farms but an increase for others, as wolf packs are simply pushed from one locality to another. In the Mercantour and Vanoise massifs, wolf kills have produced little if any change in the overall depredation rates of wild prey species and nearby livestock (Meuret and Regnier 2021).

Discussion

In both the GYE in the United States and the Provence-Alpes-Côte d'Azur and Auvergne-Rhône-Alpes regions in France, wolf restoration, and the management and policy challenges it brings, have played out in remarkably similar ways. The national parks—Yellowstone in the United States and Vanoise and Mercantour in France, served as initial hubs for wolf restoration. They provided the greatest levels of protection for the species, allowing it to gain a foothold for population growth and proliferation well beyond park borders. As places of national import and magnets for recreational tourism, the parks also provided opportunities for ecotourism and programs that enabled positive views of wolf restoration for visiting urban residents. In contrast, as national parks in both countries are also situated within largely rural areas, the rapid spread of wolves and wolf packs into areas dependent on livestock production has meant the parks have also served as focal points for conflict and opposition. In short, national parks remain central to debates over wolf restoration in both countries.

In the United States, debates over wolf reintroduction have occurred in or near Yellowstone for decades. The hunting and killing of "celebrity" wolves who stray out of the park have sparked intense public responses. In 2014, approximately 150 protestors gathered outside the gates of Yellowstone National Park in response to the wolves that were legally hunted, trapped or snared since the species in this region was removed from the federal endangered and threatened species list (Zuckerman 2014).

In France, fears over the negative impact of wolves on livestock led a group of over 50 farmers to kidnap Vanoise National Park's president, Guy Chaumeruil, and director, Emmanuel Michau, in protest. This action by farmers in the Auvergne-Rhône-Alpes region did not stand alone. Farmers throughout the entire country have made their opinions about the species' return very clear on a number of occasions. In 2014, farmers brought sheep to graze at the Eiffel Tower as they called for increased protection for their livestock (Iyengar 2014). And in 2017, farmers released their sheep into the streets of Lyon to protest wolf protections (Euronews 2017).

Nonetheless, in both countries, the general public's view of wolves has tended to be positive, even as management approaches move away from strict protection. In the United States, with the exception of the lands inside Yellowstone National Park, lethal removal to protect livestock has always been part of the restoration management plan. However, a significant change occurred with removal of wolves from the Endangered Species List and the approval of state management plans that all included some amount of hunting or culling to maintain the population at a predetermined level along with compensation programs for local ranchers. In France, the policy shift happened even more quickly, at least in terms of the scale of its application. Moreover, French policy approaches included an even greater array of tools, including preventative measures funded at the national level for defensive structures, the purchase of guard dogs and hiring of additional shepherds for livestock protection.

As Mech (2017) and others have long noted, absent a renewed national-level extirpation program, or heavy-handed (often illegal) hunting initiatives, once established, wolf populations in the United States and Europe are robust enough to continue expanding into their historic range. They are limited only by the availability (or density) of prey, which tends to be plentiful, either in terms of wildlife or domestic livestock. As wolves move ever closer to human populations—as they attempt to co-exist and share the landscape with humans—conflict is inevitable.

From a policy and management perspective then, the case of France offers lessons for the United States. From the outset, wolf restoration in France took place within a geographic context characterized by relatively higher human population densities and a higher concentration of both livestock and farming operations. Consequently, the social and economic impacts in France have been much more severe. Yet, these are precisely the types of landscapes into which wolf populations are currently expanding in the United States.

Ecologists note that opposition to wolf restoration correlates with proximity: those living closer to wolves, especially in rural areas where livelihood depends on livestock production, find themselves clashing with ever expanding wolf populations. As wolves move into urban and suburban areas then, perspectives and attitudes can quickly evolve, creating support for more flexible approaches.

A notable example in France involved the European Commission President Ursula Von der Leyen. After her pet pony, Dolly, was attacked by wolves, von der Leyen stated "The comeback of wolves is good news for biodiversity in Europe. But the concentration of wolf packs in some European regions has become a real danger especially for livestock." Following the event, the European Commission began taking action to reconsider and reduce legal protections for wolves (Barkham 2024).

In France, as in the United States, the general public's social response to wolf protection continues to be positive. Simon et al. (2023) found that 73.4% of respondents believed it important to maintain wolf populations for future generations. Nonetheless, government policies reflect a mix of pre-emptive protection measures for livestock growers, habitat buffers in national parks and protected areas, and increasing support for the lethal removal of wolves elsewhere in order to maintain populations at a prescribed number. Growing public tolerance of such measures will be vital in determining the future viability of this approach. In this way, the French experience may very well provide important signposts for the future of wolf restoration in the United States.

Declarations

Conflict of interest The authors have no conflict of interest to declare that are relevant to the content of this chapter.

References

Barkham, P. 2024. A wolf killed the EU president's precious pony—Then the fight to catch the predator began. *The Guardian* (Jan. 27). Accessed February 22, 2024. https://www.theguardian.com/environment/2024/jan/27/a-wolf-killed-the-eu-presidents-precious-pony-then-the-fight-to-catch-the-predator-began

Benhammou, F. and P. Degeorges. 2023. Macron Goes on a Wolf Hunt. *Le Monde Diplomatique* (June 1). Accessed August 10, 2023. https://mondediplo.com/2023/06/14wolves

Bernard, D. 1981. *L'Homme et Le Loup: Espace des hommes*. Paris: Berger-Levrault.

Boitani, L., M. Phillips and U. Jhala. 2018. *Canis lupus. The IUCN Red List of Threatened Species*. Updated 2023. Accessed May 12, 2024. https://www.iucnredlist.org/species/3746/247624660

CRT (Comte Regional du Trouisme). 2020. *Tourism Key Figures 2020 Edition*. Cote d'Azure, France.

DREAL (Direction Regaional de L'Environment, de L'Amenagement et du Logement). 2022. *Bilan Annuels L'Aide a la Protection*. ODS. Prefet de la Region Aubergne-Rhone-Alpes.

Duffield, J., C. Neher, D. Patterson. 2006. *Wolves and People in Yellowstone: Impacts on the Regional Economy*. Report for the Yellowstone Foundation, YNP.

Duffield, J., C. Neher, and D. Patterson. 2008. Wolf recovery in Yellowstone: Park visitor attitudes, expenditures, and economic impacts. *Yellowstone Science* 16: 20–25.

Euronews. 2017. *Sheep Take to the Streets in France as Farmers Protest Wolf Protections*. Accessed April 12, 2022. https://www.euronews.com/2017/10/10/sheep-take-the-streets-in-france-as-farmers-protest-against-wolf-protection

European Commission. 2023. *Wolves in Europe: Commission urges local authorities to make full use of existing derogations and collects data for conservation status review* (Sept. 4). Accessed November 15, 2023. https://ec.europa.eu/commission/presscorner/detail/en/ip_23_4330

FERUS (National Association for Conservation of Bear, Wolf and Lynx in France). 2019. *Loup : Pour en Finir Avec les Contre-Vérités sur le Pastoralisme et sur la Chasse*. (Mar 25). Accessed Aug. 21, 2022. https://www.ferus.fr/actualite/loup-pour-en-finir-avec-les-contre-verites-sur-le-pastoralisme-et-sur-la-chasse-2

Fogleman, V. 1989. American attitudes towards wolves: A history of misperception. *Environmental Review* 13 (1): 63–94.

Foster, D., G. Motzkin, D. Bernardos, and J. Cardoza. 2002. Wildlife dynamics in the changing new England landscape. *Journal of Biogeography* 29 (10/11): 1337–1357.

France 24. 2011. Wolf-Hunting Permit Issued after Sheep Massacre (July 25). Accessed August 10, 2023. https://www.france24.com/en/20110725-wolf-hunting-permit-issued-after-sheep-massacre-france-wildlife

French, B. 2023. Montana Wolf Harvest Drops for 2nd Year in a Row, Despite Laws to Make Hunting, Trapping Easy. *Spokesman* (March 24). https://www.spokesman.com/stories/2023/mar/26/montana-wolf-harvest-drops-for-2nd-year-in-a-row-d/

Gauvin, A. 2013. *Organization d'une Meute de Loups (Canis lupus lupus) Captifs au Parc Alpha: Impact d'un Changement de Couple Reproducteur*. Doctoral dissertation. École Nationale Vétérinaire d'Alfort.

Henshell, R. 2023. Wolf Numbers Increase in France, Rules on Hunting Them to be Relaxed. *The Connexion* (Sept. 19). Accessed February 12, 2024. https://www.connexionfrance.com/news/wolf-numbers-increase-in-france-rules-on-hunting-them-to-be-relaxed/188343

Huisman, N. 2018. *France implements new six-year wolf plan. European Wilderness Society.* Accessed February 12, 2024. https://wilderness-society.org/france-implements-new-six-year-wolf-plan/

Huisman, N. 2019. *Wolf killing continues across Europe despite its protection.* European Wilderness Society (Mar. 1). https://wilderness-society.org/brutal-wolf-killing-continues-across-europe/

Idaho Department of Fish and Game. 2023. *Idaho Gray Wolf Management Plan, 2023–2028.* Boise, ID.

ITRR (Institute for Tourism and Recreation Research) & RRC Associates. 2022. *Yellowstone Wildlife-Related Activity: Valuation Study*. Missoula, MT: University of Montana.

Iyengar, R. 2014. France: Farmers Bring Sheep to Eiffel Tower to Protest Wolf Attacks. *Time* (Nov. 28). Accessed Aug. 9, 2021. https://time.com/3609276/french-farmers-bring-sheep-to-the-eiffel-tower-to-protest-wolf-attacks/

Keiter, R. 2020. The greater Yellowstone ecosystem revisited. *University of Colorado Law Review* 91 (1): 1–182.

Lohrenz, L. 2023. *Wyoming Sheep and Lamb Losses*. National Agricultural Statistics Service, Cheyenne, WY.

Maldonado, M. 2023. Wildlife Conservation Groups Petition US Forest Service to Stop Aerial Wolf Hunting in Idaho. *Idaho Capital Sun* (December 8).

Maus, I. 2007. Regional Development and the French National Parks: The Case of Vanoise National Park. In *Protected Areas and Regional Development in Europe*, ed. I. Mose. London: Ashgate.

Mech, D.L. 2017. Where can wolves live and how can we live with them? *Biological Conservation* 210: 310–317.

Mech, D.L., and L. Boitani. 2003. *Wolves: Behavior, ecology, and conservation.* Chicago: University of Chicago Press.

Mills, K. 2023. *Wyoming Gray Wolf Monitoring and Management 2022 Annual Report*. Wyoming Game and Fish Department, U.S. Fish and Wildlife Service, National Park Service, USDA-APHIS-Wildlife Services, and Eastern Shoshone and Northern Arapahoe Tribal Fish and Game Departments. Cheyenne, WY.

Ministère de la Transition Écologique et de la Cohesion des Territories, Ministry of Agriculture and Food. 2018. *Plan National d'Actions Loup et Activités d'Élevage, 2018–2023*. Paris.

Ministère de la Transition Écologique et de la Cohesion des Territories, Ministry of Agriculture and Food. 2024. *Plan National d'Actions Loup et Activités d'Élevage, 2024–2029*. Paris.

Moriceau, J.-M. 2013. *Vivre avec le loup? Trois mille ans de conflit*. Paris: Tallandier.

Moriceau, J.-M., and E. van Maanen. 2023. A history of wolves and people in France. In *The Wolf: Culture, nature and heritage*, ed. I. Convery, O. Nevin, E. van Maanen, P. Davis, and K. Lloyd, 21–35. Rochester, NY: Boydell Press.

Meuret, M. and E. Regnier. 2021. *Wolves and Livestock Farming in France: An Assessment of 27 Years of Coexistence*. Institut National de la Recherche Agronomique. Accessed February 12, 2024. https://www.inrae.fr/en/news/wolves-and-livestock-farming-france-assessment-27-years-coexistence

News Wires. 2023. *French Plan to Manage Wolf Population Angers Farmers and Conservationists Alike*. France 24 (Sept. 18). Accessed November 15, 2023. https://www.france24.com/en/environment/20230918-french-plan-to-manage-wolf-population-angers-farmers-and-conservationists-alike

OECD (Organization for Economic Cooperation and Development). 2022. *Regional Demography*. Accessed February 10, 2023. https://stats.oecd.org/Index.aspx?DataSetCode=REGION_DEMOGR

OFB – Office Français de la Biodiversité. Bilan du Suivi Estival de la Population de Loups – Été 2021. *Le Loup en France*. Paris. https://www.leloupfrance.fr

Parks, M., K. Podruzny, S. Sells, T. Parks, N. Lance, W. Cole, T. Smucker. S. Bhattacharjee 2023. *Montana Gray Wolf Conservation and Management: 2022 Annual Report*. Montana Fish, Wildlife & Parks. Helena, MT.

Poulle, M.-L., B. Lequette, and T. Dahier. 1999. La Recolonisation des Alples Françaises par le Loup de 1992 à 1998. *Bulletin Mensuel ONCFS* 242: 4–13.

Simon, R.N., H. Fournet, N. Drouet-Hoguet. 2023. *Attitudes et perceptions à l'égard des loups et de leur conservation*. Rapport technique. OFB (Office Français de la Biodiversité) et Projet LIFE, WOLFALPS EU. 18 NAT/IT/000972.

Smith, D., R. Peterson, D. MacNulty, and M. Kohl. 2016. The big scientific debate: Trophic cascades. *Yellowstone Science* 24 (1): 70–71.

Smith, D., D. Stahler, and D. MacNulty, eds. 2020. *Yellowstone wolves: Science and discovery in the world's first national park*. Chicago: University of Chicago Press.

Sommer, E. 2021. *Montana Sheep and Lamb Losses—2020*. USDA, National Agricultural Statistics Service, Montana Field Office. Helena, MT.

Thompson, R. 1992. *Wolf-hunting in France in the reign of louis XV: The beast of Gevaudan*. Lewiston, NY: Edwin Mellen Press.

Tizzani, P., M. Bessone, L. Rossi, and P.G. Meneguz. 2022. Does predation risk affect spatial use in an introduced ungulate species? The case of a Mediterranean mouflon alpine colony. *European Journal of Wildlife Research*. https://doi.org/10.1007/s10344-022-01611-9.

US Census Bureau. 2024. *State Population Totals, 2010–2020*. Accessed February 15, 2024. https://www.census.gov/programs-surveys/popest

USDA, National Agricultural Statistics Service. 2024. *2023 State Agriculture Overview*. Accessed January 15, 2024. https://www.nass.usda.gov/Quick_Stats/Ag_Overview/stateOverview.php?state=WYOMING

U.S. Department of the Interior. 2023. *Wolf Restoration*. National Park Service. https://www.nps.gov/yell/learn/nature/wolf-restoration.htm

Van Mannen, E. 2023. The Wolf in the human mind across space and time. In *The Wolf: Culture, nature and heritage*, ed. I. Convery, O. Nevin, E. van Maanen, P. Davis, and K. Lloyd, 1–20. Rochester, NY: Boydell Press.

Wielgus, R., and K. Peebles. 2014. Effects of Wolf mortality on livestock depredations. *PLoS ONE*. https://doi.org/10.1371/journal.pone.0113505.

Wilson, R.K. 2020. *America's public lands: From Yellowstone to Smokey bear and beyond*, 2nd ed. Lanham, MD: Rowman and Littlefield Publishers.

Yin. 2017. *Case study research and applications: Design and method*, 6th ed. Thousand Oaks, CA: Sage.

Young, S., and E. Goldman. 1964. *The wolves of North America*, vol. 1. Mineola, NY: Dover.

Zuckerman, L. 2014. Conservationists rally outside Yellowstone to protest wolf hunting. *Reuters*. https://www.reuters.com/article/us-usa-wolves-rally/conservationists-rally-outside-yellowstone-to-protest-wolf-hunting-idUSKBN0F400620140629

CHAPTER 6

Failed National Park Proposals in "Sweet Home Alabama"

Lary M. Dilsaver and Joe Weber

Among the U.S. federal land agencies, the National Park Service (NPS) has the sternest legal duty to preserve the natural and historical resources of the country while providing education and inspiration to the people. This mandate, now firmly part of the agency's culture, is only exceeded by designated wilderness areas some of which are within national parks (Dilsaver 2016, 34, 245–54). The units in the national park system must preserve and exhibit compelling resources of interest to the entire country's people. Many proposed areas fail to pass inspection due to a lack of this "national significance" or for other problematic qualities. The following essay analyzes that procedure in one state—Alabama—to

L. M. Dilsaver (✉)
Department of Earth Sciences, University of South Alabama, Mobile, AL, 36688, USA
e-mail: ldilsaver@southalabama.edu

J. Weber
Department of Geography, University of Alabama, Tuscaloosa, AL 35401, USA
e-mail: jweber2@ua.edu

J. Weber and S. Sultana (eds.), *The Changing Geography of National Parks and Protected Areas*,
https://doi.org/10.1007/978-3-031-74653-6_6

explain the review process, the decisions made by the NPS, and what they reveal about the culture of preservation and the agency's political and procedural agendas.

After summarizing the growth of the national park system and explaining how the NPS evaluates a proposal, we introduce the status of federal and state parks in Alabama and list 72 places that failed to be added to the list of national units. We break down the proposed sites by type and then answer four questions: Who proposed them? Why did they fail? What subsequently happened to them? And what does this mean for the preservation of culture in the state and nation? We append a list of the documents and files from which we drew the names and data on the 72 failed proposals.

The first national park in the United States, and in the world, was Yellowstone established on March 1, 1872. Thirty-four more parks and monuments followed before Congress passed the National Park Service Act on August 25, 1916. This "Organic Act" organized the extant parks and monuments into a system and brought order to the process of identifying potential additions. Congress and most presidents have added hundreds of new units while facing a continuous stream of proposals for new units from political officials and private citizens. In March 2024 the American national park system consisted of 429 units spread across all 50 states and most territories.

In Alabama, the national park system roster includes six historic sites, two recreation corridors, and one natural preserve (Fig. 6.1). Five of them were established after 1982. Alabama's total of nine is roughly the average number of units for each state in the country. At the next level of preservation, Alabama has 25 state parks and the Alabama Historical Commission (AHC) manages another 14 structures (Encyclopedia of Alabama 2022). This might lead to a comfortable assumption that the state's heritage is well preserved. Does this reflect a commitment to preservation among Alabamians in environmental and historic preservation? What other Alabama sites might have been included in the national park system to educate and inspire the populace? The answer to this question can be found by looking more deeply into the NPS experience in the state. Gaining inclusion in the elite national park system is difficult and each state has fielded more proposals that failed than ones that succeeded. Our research in Park Service archives has turned up records for 72 sites in Alabama that somebody thought worthwhile for the national system, but which failed to make the grade (Fig. 6.2). Analysis of these proposals

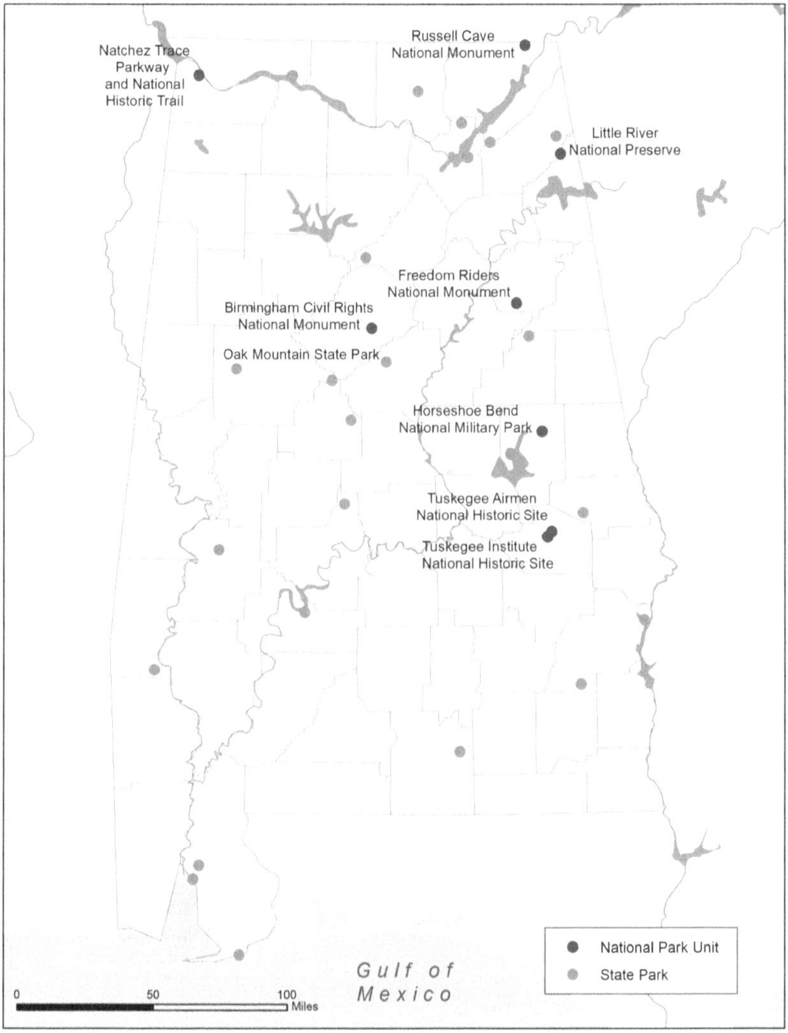

Fig. 6.1 State and national parks in Alabama. *Source* Authors

illuminates much about the natural and historical resources in the state, why so many proposals failed, and what became of them.

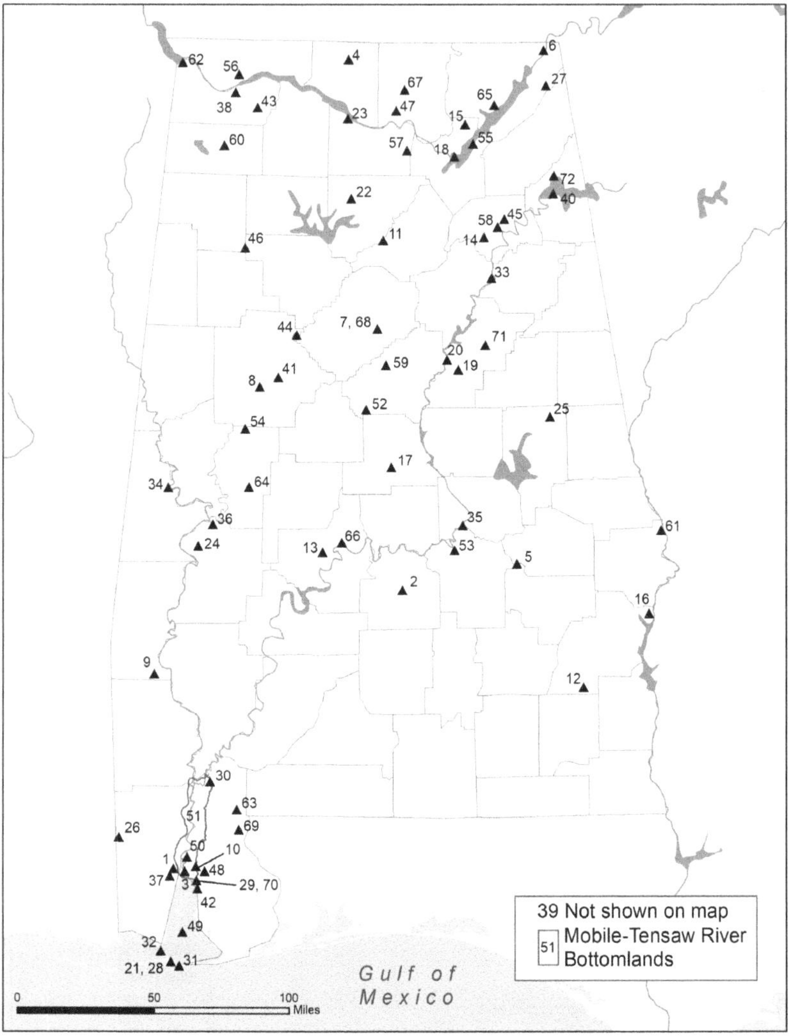

Fig. 6.2 Proposed national parks that failed in Alabama. *Source* Authors

The History of Park-Making

The history of post-1916 park-making in the United States is divided into four periods. The earliest spanned the years from 1916 to the end of World War II. NPS Director Stephen Mather's vigorous campaign to popularize the parks led government officials and ordinary citizens to propose many new sites across the nation. Park Service officials deemed many of these proposals either lacking in importance to the entire country or excessively focused on outdoor games and recreation. Mather co-hosted the country's first conference on state parks in 1921, hoping to foster secondary systems to meet the recreation needs of the country (Conard 1997, 28–43) During the 1930s, when Works Progress Administration and Civilian Conservation Corps funds and labor were available, Park Service planners helped state and local officials develop hundreds of parks that did not qualify for the national park system. Ironically, the National Park Service's leadership in state and local recreation planning ultimately led Congress to order the agency to identify outdoor recreation sites for the national system (Unrau and Williss 1983, 106–28).

The Park Service also developed a system of natural and historic themes to guide the search for places that best represented each scientific or historical category. Proposing a unit that filled a niche in the thematic framework let the NPS justify it to Congress. Director Horace Albright, lead field investigator Roger Toll, and anthropologist Clark Wissler helped design the first set of historic themes, while geographer Nevin Fenneman's map of physiographic regions in the United States became an analytical tool for measuring the worth of natural resource sites. Although these evaluative categories were in use by the late 1920s, the Park Service issued its first full thematic framework in 1936 (Fenneman 1916, 261–353; Wirth 1980, 163–66). With Depression-era funds and its thematic framework in hand, the National Park Service undertook large-scale studies with the expressed purpose of identifying the best of each category. Nearly a third of Alabama's park proposals came during the 1930s.

The end of World War II initiated a second period during which meager funding slowed the addition of new parks to the system. Nevertheless, the Park Service continued to study both new and repeat proposals, often with private financial backing. The importance of filling out a representative system of resources grew as funds remained limited. Although Mission 66 temporarily boosted appropriations, the bulk of the money went to new administrative and visitor infrastructure. Criticism of Director Conrad Wirth's focus on development coincided with a surge of environmentalism and with the release of the Outdoor Recreation Resources Review Commission's reports. The latter spurred more interest in recreation areas, rivers, trails, and parkways (Wirth 1980, 237–84; Carr 2007, 289–338).

The combination of these upheavals in the National Park Service's focus and management led to the third period of park-making from the early 1960s through 1980. The agency issued new appraisals of the national park system thematic framework and its quotient of representation in 1972. In the natural history plan the Park Service divided the country into 41 natural regions based on Fenneman's physiographic map and devised 31 geological and ecological themes that meshed to create a large matrix of almost 450 distinct resource categories. Only half of these were represented by the existing units within the park system (USDI 1972b).

The complexity of the history plan meant that the Park Service and Congress had much more work to do to represent the nation's cultural heritage. The plan highlighted nine major themes that were subdivided into 43 subthemes. These subthemes in turn were further divided into 281 historical "facets" that needed to be represented if the park system were to be "truly representative of the country's national heritage." NPS officials determined that the historic parks then in the system represented only 32% of the identified facets (USDI 1972a). Congress then passed P.L. 94–458 in 1976 which mandated that the Park Service use Land and Water Conservation Fund money to propose at least twelve new parks per year. Overall, the years of President Jimmy Carter's administration rivaled the 1930s for the number and variety of new parks proposed.

The arrival of the Reagan administration in 1981 curtailed this bonanza of park proposals and ushered in the fourth period of park-making. Budget-conscious legislators ordered Park Service planners to carry out only proposed park studies authorized by Congress in 1998 (Public Law 105–391). In response, the Park Service produced a new set of thematic plans that reflected a more realistic appraisal of the existing units of the national park system plus two decades of new parks added specifically to address gaps in the system. The 1990 natural plan shows that nearly all the natural resource themes are now included in the system. The agency's 1996 history framework focuses on thematic gaps in the system and stresses their representation within existing parks wherever possible. The historical parks authorized since the early 1980s include the history of women, minorities, and themes like labor history, arts and literature, and architecture (NPS 1990).

To determine a site's worthiness the NPS developed three basic criteria for judging proposed units. The first and foremost is "national significance." Though qualitative and subjective, the question that a new area must satisfy is, "How important is the resource to the entire nation?" This includes both its level of interest to the American people and whether it is the best representation of a historic or natural resource theme. Field investigation teams are staffed with senior Park Service specialists to reduce subjectivity as much as possible. Agency experience has always counted heavily in proposed area surveys. When ordered to evaluate a potential national seashore, one field team member asked for further explanation of the concept. His supervisor responded that he would know it when he saw it (Everhart 1998). Sites that are "not of national significance" are rejected as unqualified.

The second criterion is feasibility. Regardless of its significance, a proposed park that will cost too much for land acquisition or meet intense local opposition is not considered feasible. The third criterion is suitability. A proposed park should not duplicate resources already in the park system. In addition, a proposed park is not deemed suitable if there is a question about the historical or ecological integrity of the site or if the site is not accessible to the public. In recent years, a de facto fourth criterion has been added by Congress: Can the resource be protected and available to the public by another government agency or private organization? Budget-conscious legislators worry about the costs of an expanding park system (Smith 2003).

Alabama's National Park Units

The nine NPS units in Alabama include one natural preserve (Little River Canyon National Preserve), six historic sites (Birmingham Civil Rights NM, Freedom Riders NM, Horseshoe Bend NMP, Russell Cave NM, Tuskegee Airman NHS, Tuskegee Institute NHS), and two recreation corridors (Natchez Trace National Trail and Natchez Trace National Parkway). Five were established after 1982. The 72 failed sites cover a wide variety of places and purposes. Analysis of the fates of these proposals illuminates what cultural, political, or economic elements led to their disapproval. Five questions can be asked to better understand the history of Alabama's proposed parks: (1) What kinds of areas have been proposed within the state? (2) Who proposed these areas for national park status? (3) Why did the NPS reject or ignore these areas? (4) What has happened to these sites? (5) What do these failed parks and their ultimate fates tell us about how preservation works in Alabama and the United States?

What Kinds of Areas Have Been Proposed in Alabama?

NPS units preserve three types of resources—natural, historical, and recreational. Many parks contain all three types of resources, but nearly all have a dominant purpose that led to their establishment. Like most eastern states, Alabama's 44 proposed historical sites account for the majority (61%) of all submissions. This is close to the ratio of six historical NPS units among the state's total of nine. Eighteen areas nominated for recreational use make up another 25%. Ten natural resource areas or 14% are in the smallest category, although they include what many argue is the single most important resource in the state. Most proposals were made during the early years from 1916 through the end of World War II in 1945. The NPS fielded 53% of all the failed proposals in Alabama, while only the multistate Natchez Trace National Parkway joined the system during this period. Many of the failures were repeatedly nominated without success.

Historical Sites

Five types of historical sites explain the influential events of the state—Native American vs. Euro-American battlefields, Civil War forts and battlefields, areas important to minority peoples, places with economic development stories, and sites associated with extraordinary individuals. Two wars that took place in Alabama dominate the state's historical memory—the United States v. Creek War and the Civil War. Early in the nineteenth century, a Red Stick Creek contingent, led by William Weatherford, fought to maintain their land and way of life under pressure from white settlers. On August 30, 1813, a force of 700 Red Sticks, attacked 400 Americans, Indian allies including other Creeks, and slaves at a fortified plantation called Fort Mims. Over several hours they massacred some 250 men, women, and children. The outrage felt by white frontier settlers led Andrew Jackson and volunteers from Tennessee and elsewhere to engage in a string of battles at Emuckfaw, Fort Strother, Fort Tombecbee,

Talladega, and finally Horseshoe Bend. The last of these destroyed the Red Sticks and forced the Creeks to cede over 21 million acres (8,498,398 hectares) of land. The treaty was signed at Fort Toulouse which was located on the ruins of an earlier French fort. All these locations were nominated as potential park units, but only Horseshoe Bend, a part of the national park system since 1956, was successful. The Park Service judged the other proposals not feasible because they lacked the remaining infrastructure to interpret, not suitable because Horseshoe Bend was so much more important, or in the case of Fort Toulouse, not of national significance (Encyclopedia of Alabama 2022).

An earlier fight involving Europeans and Native Americans occurred on October 9, 1540, when Spanish troops exploring what are now the southern U.S. states encountered Chief Tuscaloosa at his fortified village called Mabila (or Maubila). Hernando de Soto landed near Tampa, Florida in 1539 with approximately 600 men, plus weapons, horses, and supplies. Searching for gold, pearls, and other treasure while demanding food, porters, and "comfort women" for his soldiers, the conquistador traveled a long path through Georgia, the Carolinas, and Tennessee, before crossing Alabama (Fig. 6.3). Tuscaloosa and his warriors knew what to expect and attacked some of the visiting Spanish inside the stockade. A massive battle ensued which killed 20 Spanish troops and injured more than 150. They also lost most of their equipment including religious materials. Nevertheless, the armored and horse-riding Spaniards killed approximately 2500 Indians. It stands as the worst defeat for Native Americans in U.S. history. It also broke the Spanish expedition and led to De Soto's death near the Mississippi River (Knight 2009).

The episode has led other explorers, historians, geographers, and archaeologists to embark on a centuries-long search for the exact site of Mabila. No trace was found by two later sixteenth-century Spanish expeditions, and the location remains unknown. Vernon Knight (2009) called it the Holy Grail of archaeological mysteries in the Southeast. Recent research by archaeologist Ashley Dumas along the border between Marengo and Sumter Counties has uncovered multiple pottery sherds and bits of Spanish metal from that period, but no definite trace of a burned stockade (Tutor 2024). The hunt will continue.

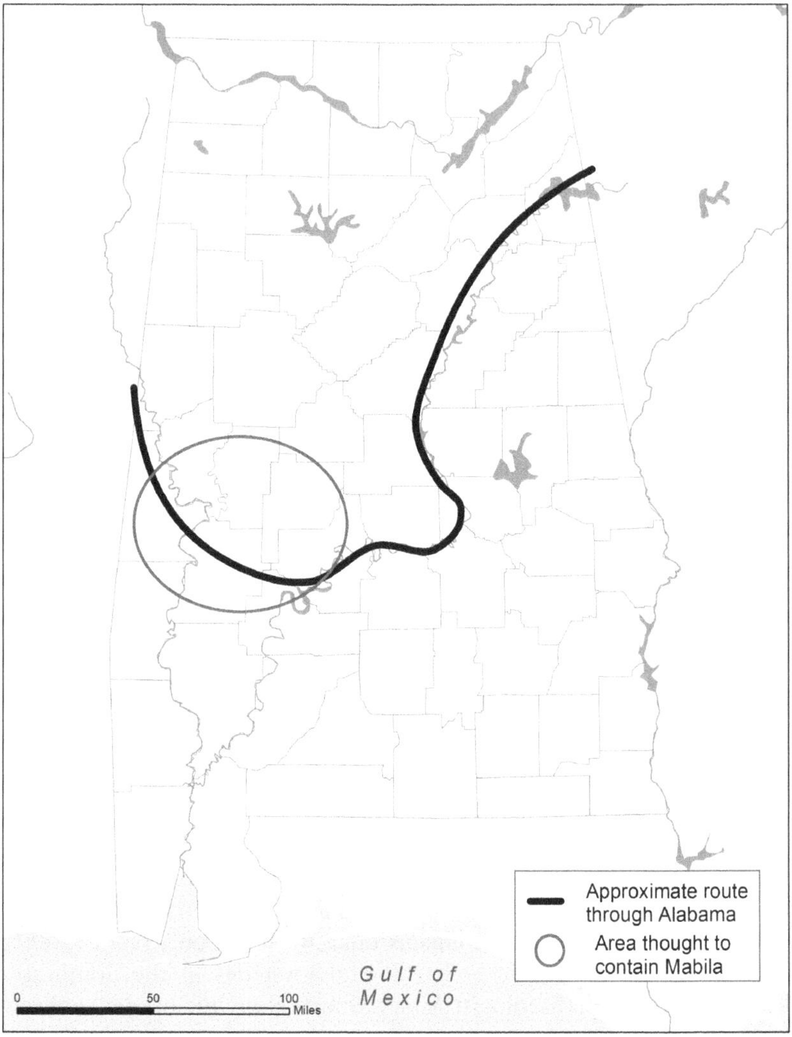

Fig. 6.3 De Soto's route and the area where chief Tuscaloosa resided at Mabila. *Source* Authors

Civil War Sites

If there is one event that still lives in the minds of many people in Alabama it is the Civil War. Fifteen battle sites and forts make up 21% of the total proposals in the state. The majority were proposed before World War II showing a decline as the people who remembered it passed away. Some occurred in the northeastern part of the state, but the most important Civil War battles in Alabama took place in and around Mobile Bay. The city of Mobile was one of the last open ports of the Confederacy. Union forces hoped to shut down trade and further cripple the Southern economy. In addition, President Abraham Lincoln was running for reelection and faced challenges due to war-weariness. A critical victory was necessary to improve morale. On August 5, 1864, Admiral David Farragut led four ironclads and more than a dozen wooden ships into the bay uttering the famous command, "Damn the torpedoes [mines], full speed ahead!" At the same time Union troops attacked Fort Gaines on Dauphin Island, a ramshackle earthen Fort Powell on the mainland north of the island (Fig. 6.4), and Fort Morgan on the eastern peninsula across the mouth of the bay. Once in the bay and having lost the ironclad USS Tecumseh to a "torpedo," Farragut's fleet met a Confederate ironclad and a few other ships. Three hours of pounding finally forced the armored CSS Tennessee to surrender. A few days later the forts at the mouth of the bay also surrendered (Fig. 6.5). Farragut's victory in Mobile Bay was one of the badly needed Union victories that ensured President Abraham Lincoln's reelection in November (American Battlefield Trust 2024).

The land campaign to take Mobile occurred from April 2–9, 1865 as Union commanders looked to wipe out remaining Confederate forces in the deep south and stop their production of Confederate arms. Union troops moved north from the coastal forts along the eastern shore of Mobile Bay toward Confederate defenders at Spanish Fort, arriving there on March 27. CSA forces only numbered about 3000 men but had artillery firing on Union gunboats from a hill called Fort McDermott (Fig. 6.6). They held out against the much larger Union army until the Spanish Fort fell on April 8. Union General Edward Canby concentrated 16,000 men for the attack on nearby Fort Blakeley the following day. After a brief fight, Union troops breached the Confederate breastworks and trenches compelling the defenders to surrender. As the victorious troops approached Mobile from the east, city fathers surrendered to avoid

Fig. 6.4 Fort gaines in 1935 when the NPS investigated it. *Source* NPS—HABS/HAER files

destruction. Around 5000 African Americans played a major role in the successful Union assault. The capture of Fort Blakeley was the last important battle of the war, fought the same day Robert E. Lee surrendered his army at Appomattox (American Battlefield Trust 2024).

All the sites named here were proposed as units of the national park system, including the individual-named forts as well as the naval battle in the Bay. Had these sites been added they would have been a unique collection in the park system, focusing on the Bay rather than a single battlefield.

Many other small skirmishes were fought in northern and eastern Alabama, but the NPS rated all the Civil War battle sites including those at Athens, Battery Hill, Days Gap, Decatur, Lookout Mountain, Montgomery, Phoenix City, and Selma as either not of national significance or not suitable because of better examples in the system. One last proposed war-related site was 37,000-acre (14,973 hectares) Camp Sibert, a World War II chemical warfare training base in Etowah and St. Clair Counties.

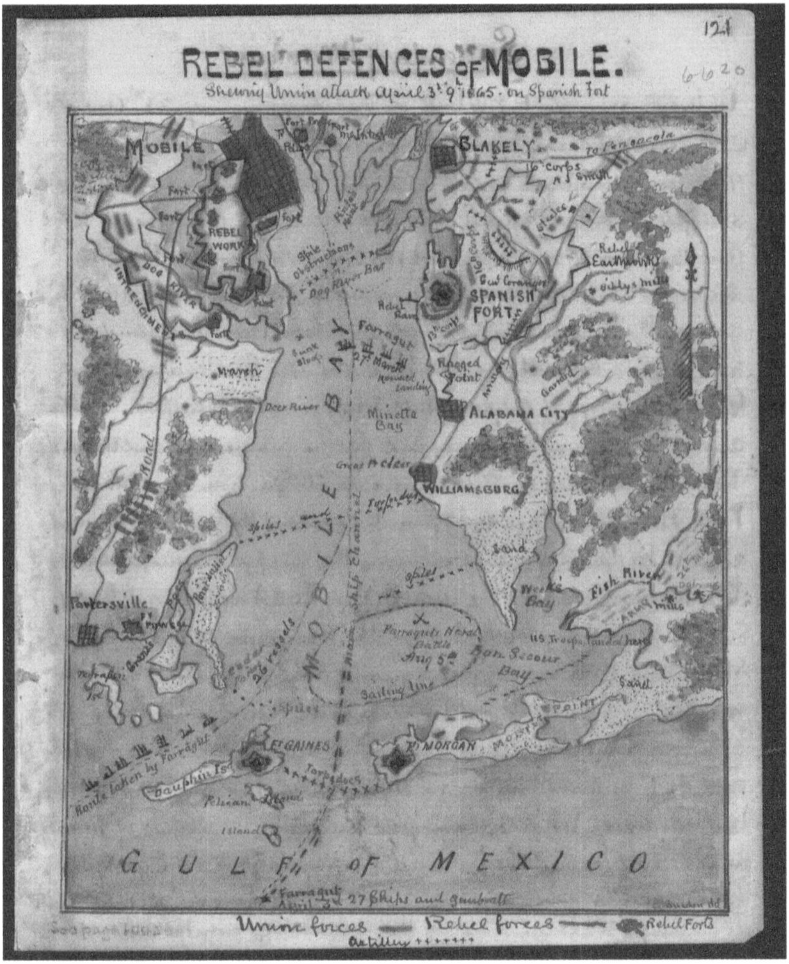

Fig. 6.5 An 1865 map of the mobile bay civil war battles. Drawn by Robert Sneden. *Source* Library of congress

Mayor J. H. Meighan of Gadsden proposed it with support from Senator Lister Hill. The NPS inspected it in 1945 and found it not suitable, presumably because of the traces of its former activity.

Fig. 6.6 Fort McDermott in Spanish Fort. Photo by Lary Dilsaver

Sites Commemorating Black and Native American History

In 2019 divers discovered the remains of the *Clotilda*, the last ship to carry African slaves to the United States. It was deliberately sunk in the Mobile River by its owners after they illegally smuggled the captives into the area in 1860. The Africans settled near the town of Prichard just north of Mobile on land that became an industrial zone. At a hearing in April 1983, Mayor John H. Smith of Prichard sought historic preservation for the community and memorialization of the plight of the Africans and their descendants. However, the Park Service responded that there were no "site, buildings, or objects of national significance" and that the area had no historical integrity. Hence, it lacked suitability. Subsequent discovery of the ship's burned timbers has not renewed the proposal.

The NPS files list two other proposals for African American sites. A multistate proposal for Historical Black Colleges is a funding program that does not propose a specific place for the system. The Alabama Black Belt National Heritage Area is an enormous swath across the middle of the state where rich soils led to a plantation economy that relied upon Black workers. The region still has a substantial population of their descendants and many civil rights events have occurred there. National heritage areas are large regions filled with many private and other government components that are managed by committees of stakeholders including the NPS.

The agency considered three archaeological sites representing Native Americans other than existing system units Russell Cave and Horseshoe Bend. South of Tuscaloosa is Moundville managed by the University of Alabama (Fig. 6.7). Located on the Black Warrior River, the park preserves an 800-year-old, 326-acre (132-hectare) village site where Mississippian people constructed 29 massive flat-topped earthen mounds surrounding a large central plaza. They served as elevated platforms for civic and ceremonial structures and the homes of nobles. The state acquired the property in 1933 before the NPS could act. Eight years later, after remedial work by the Civilian Conservation Corps, the NPS Advisory Board, created by the Historic Sites Act of 1935, declared that it was nationally significant. The state did not release it. In the delta, at the head of Mobile Bay, is a smaller set of 18 mounds. Hazel Council, a member of the Alabama Federation of Women's Clubs, partially owned the property and proposed it in 1939. The Park Service declared that it was not feasible due to a lack of funds for its purchase. The third historical native site lay on the Coosa River close to the purported De Soto Trail. Despite support from Senator Hugo Black, the NPS deemed it not to have national significance in 1935.

Sites of Economic Importance

In addition to the Alabama Black Belt National Heritage Area, another heritage area at Muscle Shoals and one still under consideration along the Chattahoochee River were studied by the NPS. NHAs are not technically part of the park system although the agency proposes, studies, and co-manages them. The NPS investigated a third one based on 600 coal and steel sites in and around Birmingham in 1991 and found it to be not suitable. Seven other places of economic or transportation renown included two plantation houses, Gaineswood in Demopolis (Fig. 6.8) and the home of naval hero, congressman, and prohibitionist Richmond P. Hobson in Greensboro; industrial complexes Sloss Furnace in Birmingham (Fig. 6.9) and Old Cedar Creek, the first iron mill in the state;

Fig. 6.7 A portion of the Moundville site. *Source* The George F. Landegger collection of Alabama photographs in Carol M. Highsmith's America, library of congress, prints and photographs division

the remains of the Tennessee-Alabama-Georgia Railway; NASA's Man in Space facilities at Huntsville; and finally LaGrange College, the first site of higher education in the state now controlled by its descendant the University of North Alabama. The National Park Service found Gainesville not suitable and all the rest not of national significance.

Historical Individuals

Four sites celebrate extraordinary individuals who lived or acted within Alabama. The earliest was naturalist and artist William Bartram who traversed much of the South collecting biological data and specimens in the mid-1770s. He began in Philadelphia and journeyed for 4 years to reach the Mississippi River. His work informed the world about the ecological diversity and richness of the region. He also contacted and gathered data about the Cherokee, Creek, and Choctaw peoples. His pathway is well known and portions including a canoe trail in

Fig. 6.8 Gaineswood plantation home. *Source* NPS—HABS/HAER files

Alabama celebrate his accomplishments. Local Bartram Trail Groups worked through the Bartram Trail Conference to push the government to investigate the matter. Public Law 94-527 in 1976 ordered the NPS to study it along with seven other candidates for addition to the National Trails System. Although the agency admitted that a few segments were worthwhile, it found the entire proposal not feasible.

A second heroic person was General Andrew Jackson who not only fought Red Stick Creeks across the state but passed near Mobile Bay on his way to face the British in New Orleans. Allegedly he stopped in 1814 to address his troops under a massive oak tree on the eastern shore of the bay. Henceforth it has been known as the Jackson Oak. The NPS in the mid-1930s and others have wondered if the story is true and the agency declared that it did not have national significance.

Chronologically, the third was CSA General Braxton Bragg who led armies to one dubious and heavily criticized win at Chickamauga and

Fig. 6.9 Sloss furnaces near Birmingham. *Source* The George F. Landegger collection of Alabama photographs in Carol M. Highsmith's America, library of congress, prints and photographs division

many defeats before and after that. Ultimately, Jefferson Davis relieved him of command. He was roundly hated by other officers and his men who suffered inordinate casualties in those battles. After the war, the North Carolina native moved to Mobile where his brother owned a house. He left after a few months and the home was later bought by the Mitchell family. The 1970 proposal suggested the architecture of the now-named Bragg-Mitchell House might be worth an addition to the system. The NPS understandably disagreed and noted its lack of significance.

The final notable figure was blind and deaf Helen Keller whose fame rests on her conquest of these disabilities and subsequent social and political action. Famous for her own writings and the film "*The Miracle Worker,*" she championed efforts to improve the treatment of the deaf and the blind and co-founded the American Civil Liberties Union. Congressman Robert E. Jones proposed the house in 1949 after the town of Tuscumbia took over its upkeep (Fig. 6.10). The NPS Advisory Board turned it down because the policy is not to establish a site until the

Fig. 6.10 The home of Helen Keller in Tuscumbia. Photo by Jennifer Lee. *Source* Wikipedia. Rescaled to 300dpi. https://creativecommons.org/licenses/by/2.0/deed.en

person has been dead for 25 years. Keller died in 1968 but no action has followed.

Recreation Sites

The history of Alabama's state parks began in 1927, when the Alabama legislature passed the State Land Act which ordered the development of state parks and the administration of other state lands. In 1939, a newly established Alabama Department of Conservation established Cheaha, the first state park, near the south end of the Appalachian Mountains. The goal in acquiring and establishing these parks was to "preserve wild areas and provide recreational opportunities." Activities such as camping, boating, fishing, and hiking are encouraged and everything from hotels to golf courses can be found in some of them. Many are smaller than two square miles (518 hectares) and were strategically placed near population centers. Alabamians and their legislators suggested 17 recreation spots for the national park system. Most were water-oriented occurring on rivers (Black Warrior, Cahaba, Connors Island, Coosa, Escatawpa, Hog Island, Lake Bankhead, Murphy Hill, Pickwick Lake, and Soldier Creek), beside springs (Blount and Blue), or along the coast (Dauphin Island). Two proposals sought full national park status for larger areas including Chilton County and Oak Mountain/Shelby County. Most of the rest were small places that the state wanted to fill out its recreation matrix for locals (Alabama Department of Conservation and Natural Resources 2014).

Natural Resource Sites

The Mobile-Tensaw River Bottomlands, established in 1984 as a National Natural Landmark, has had the fullest support of the NPS and can easily meet the criterion of national significance as a national park or preserve. In 1979 the agency conducted a study to determine what options for its management were available. A cost/benefit analysis rated it second highest as an NPS unit. The analysts admitted that it was for economic savings that it did not receive the highest overall score. Since that study, the proposal has appeared consistently on nearly every list submitted by the NPS with its highest recommendation. It is a large area stretching from the confluence of the Alabama and Tombigbee Rivers through the distributaries of the two named rivers to Mobile Bay. Harvard University entomologist and Pulitzer Prize winner Edward. O. Wilson described the Mobile-Tensaw River Bottomlands while participating in a recent campaign to make it a national park:

The greater Mobile-Tensaw River area, containing the Delta and the Red Hills immediately to the north, is both uniquely beautiful and one of the least known natural areas in America. Although located on the flat Gulf of Mexico plain, its habitats match—or exceed—in diversity those in any of America's established national parks...Because of the great variety of the habitats in which they live, and their still mostly undisturbed condition, the fauna and flora of the greater Mobile-Tensaw River area may well be the most diverse of comparable geographical areas in North America. The 28 + species of oaks in the Red Hills, for example, is evidently a national record. The number of turtle species is one of the highest in the world. The full roster of plants and animals includes many endangered species. Others are certain to exist still unknown to science (Waselkov et al. 2016, xix).

Chief Wildlife Biologist Greg Plumb (Waselkov et al. 2016, 7, 9) from the NPS Biological Resources Division, who co-edited a second much larger NPS study of the area in 2016, added:

It has been over 40 years since the Mobile-Tensaw River Bottomlands was championed by Alabama citizens and scientists as an area of national significance leading to National Natural Landmark designation. Since this designation, Alabama conservation agencies, Alabama academic scientists and scholars, Alabama public and private conservationists, and popular media have continued to highlight the extraordinary values and characteristics of the Mobile-Tensaw River Bottomlands as a place of local, regional, national, and international significance... It is an important habitat for wildlife, and the vastness of the delta and its limited accessibility continues to support its basic integrity.

In the same 2016 document, environmental reporter Ben Raines (Waselkov et al. 2016, 151–54) pointed out that the state leads the nation in aquatic diversity, but "ranks as the king of extinctions for aquatic creatures, responsible for roughly half of all extinctions that have occurred in the continental US since the 1800s." He explained further,

Obviously, a state cannot forever wear both the crown for most aquatic diversity and the crown for most extinctions. Unfortunately, the current political climate in the state suggests the scale may be tipping toward more extinctions. The Alabama Legislature has provided zero funding to the state's environmental agency, the Alabama Department of Environmental Management, for the last several years. And Alabama ranks dead last among the 50 states in terms of money spent on environmental protection.

So, what happened? By 2018 Alabama congressmen and senators had backed away from support for the initiative and it quietly disappeared from the legislative scene. Some portions of the area are protected by state agencies and The Nature Conservancy (Fig. 6.11).

Other natural resource sites include Noccalula Falls, four underground sites (Blount Springs Caverns, Cathedral Caverns, Shelta Cave, and Newsome Sinks Karst Area), Mallard Creek, the Splinter Hill Pitcher Plant Bog (Fig. 6.12), and two 1930s proposals for large national parks named Alabama Forest north of Mobile and Scottsboro or Crockett and Jackson, west of Huntsville. The last two vaguely bounded tracts fell to advancing settlement and economic development.

Fig. 6.11 A quiet stretch of the mobile-Tensaw rivers bottomlands. *Source* The George F. Landegger collection of Alabama photographs in Carol M. Highsmith's America, library of congress, prints and photographs division (DLC/PP-2010:090)

Who Proposed These Sites for the National Park System?

NPS files often list the identities of those who proposed new units, but many of them are the last of a long trail of unnamed enthusiasts. In seven files where only the site's name is supplied, the proposers remain unknown. The people who proposed new national park system units did so as individuals or members of private organizations. Local citizens or interest groups suggested 20 of them. Another 11 came from Alabama members of Congress, no doubt responding to such unnamed citizens or interest groups. Among identified sources, chambers of commerce suggested 10 places, various levels of state or local government eight, scientists or historians three, the official Civil War Commission three, congressional votes four, and the NPS six. The 1930s saw a vigorous cluster of proposals including ten by Mobile's Chamber of Commerce or city commissioners. Chamber members John Glennon or Herbert Brown

Fig. 6.12 Splinter Hill Pitcher plant bog in Northern Baldwin county. Photo by Robin Dilsaver

signed eight of the initiatives and Congressman Frank Boykin (in office 1935–1963) added his approval to seven of them. Most were Civil War sites around the bay. Senator Hugo Black (serving from 1927 to 1937) and Congressman Tom Bevill (in office 1967–1997) particularly sought approval for recreation sites. Bevill sponsored a meeting between the NPS, the Tennessee Valley Authority, and the AHC to investigate six places in northern Alabama including Cathedral Caverns, Connors Island, Coosa River, Hog Mountain, Murphy Hill, the Tennessee-Alabama-Georgia Railway, and Little River Canyon. The latter became an NPS unit in 1992 while all the others failed. The sequence of actions behind a proposal by the NPS to investigate a site can also remain unknown. Park units in the system usually identify the backers of its campaign for establishment in their administrative histories, but there is no reason for the details of those that fail to remain in the files.

Why Did the Park Service Reject These Areas?

The NPS cited all four criteria for judging national park proposals to reject the 72 failed sites. The agency ranked 43 of the 72 as not of national significance. Many were recreation sites. The Alabama Department of

Conservation and Natural Resources notes that more than 80% of the money to finance its state parks comes from visitor use fees, donations from chambers of commerce, civic organizations, city and county governments, cigarette taxes, and federal funds. The low amount provided by the state government is a point of pride. This explains the paucity of state parks and why other sites are sought by the national system to increase recreation opportunities without incurring more expense (Encyclopedia of Alabama 2022).

The NPS declared 23 places not feasible due to lack of funds for acquisition, intense opposition, or a dearth of public and congressional support. The Escatawpa River flows from southeast Mississippi to Mobile Bay and was proposed for the Wild and Scenic River category. People living along its course aggressively opposed this as did the governors of both states. Even when the Park Service first welcomes a proposal, an inspection can reveal problems. Mobile County's Dauphin Island faced this in the 1930s and again in the late 1960s. The NPS first rebuffed the barrier island as a recreation area while supporting one for Santa Rosa Island in nearby Florida. Later during the successful campaign to establish Gulf Islands National Seashore, it was part of the chain of barrier islands stretching from Florida to Cat Island in Mississippi. Settlement had covered the eastern half of Dauphin Island, but kilometers of low beaches lay westward. Nevertheless, the NPS abruptly dropped it from the plan not long before the seashore's authorization. Areas described as not suitable either duplicated existing resources in the system or were more appropriate as national heritage areas with their complex patterns of ownership and management.

What Happened to These Areas?

The overwhelming majority of Alabama's proposed places remain in private ownership or are a mix of private and local or state government holdings. Thirty-four are entirely private. Bladon Springs, Blakeley, Cathedral Caverns, Fort Morgan, and Oak Mountain are state parks while seven others are official AHC sites. Four are parts of local parks. Jackson Oak is in Daphne's Village Point Park Preserve while Sloss Furnace is in Birmingham, Montevallo in the town of the same name, and Noccolula Falls in Gadsden. A few other attractions such as the Helen Keller Home are privately managed but receive local or state government financial assistance. The protection level for all these places varies from excellent in the larger state parks to good in the eight that are recognized as national natural or historical landmarks or listed on the National Register of Historic Places. Thereafter the picture becomes unclear to the poor. The Mobile Bay Naval Battle and nearby Fort Powell are underwater and have no hope of in-place recognition.

Conclusion: What Are the Key Lessons from Alabama's 72 Failed Parks?

We have summarized the growth of the NPS and its means of adding new units to the system, we have reported what areas were proposed, who did so, why they failed and what happened to most of them. During the research for this article, some people have questioned the authors' motive for focusing on places that did not pass the test to be national park units. What can we learn from these stories of failure? In business, marriages, and most other human activities a lot can be learned from analyzing failure. The data in Alabama adds to a growing understanding of how preservation works in the United States. Two stories are evident from the analysis of Alabama's park proposals. First, most of the nominated sites judged by NPS standards were not of national significance. Is Alabama unusual in this pattern? Earlier work done on California and Montana indicates it is not. Both of those states had five to seven times as many failed proposals as the ones that joined the system (Dilsaver 2008; Dilsaver and Wyckoff 2009). When the natural resources of Alabama are compared to the rivers, waterfalls, caves, and forests of the national park system they are clearly inferior. Although a lot of history has happened in the state, the level of importance for most could not overcome the agency's suitability or feasibility criteria. Many of this country's 7000 Civil War battles and skirmishes took place in the state, but only the Naval Battle of Mobile Bay came close to the significance level of Gettysburg, Vicksburg, Antietam, and a half dozen others. Multiple battle sites where Euro-Americans and Native Americans fought were either peripheral to Horseshoe Bend National Military Park or, in the case of Fort Mims, do not have any structural remains to interpret. Many other historical sites were and are significant to the state but cannot compete with the compelling impact of places in other regions that draw visitors from across the nation. The problem for Alabama and for the thousands of proposals in every other state is that the national park system has the highest possible membership bar to cross.

The other story is determining whether something in the culture of Alabama's people and government affected the success rate. Was it a lack of motivation among the public and politicians to aggressively campaign for truly worthy places? Environmentalists might say yes. The Mobile-Tensaw River Bottomlands is an example of an overqualified natural resource being ignored due to an unwillingness to lose private usage opportunities and a lack of commitment by elected officials. The preservation statistics Ben Raines brandishes are factual. Some people hold a disdain for federal government influence. Alternatively, some of the Alabama places proposed to the NPS still draw interest, have public support, and would be outstanding parks or monuments. Three come immediately to mind. The Mobile-Tensaw River Bottomlands should not just slip into back files. It is a site of international importance, and the United States will fail in its environmental mission if it does not give it the highest protection. The Mabila site on the trail of De Soto should be included once it is located. The biggest-ever battle between Europeans and Native Americans is a critical historical resource. A third site surrounds Mobile Bay. Every proposed fort or battlefield in the area failed, but the naval battle was decisive militarily and politically. The NPS needs to select somewhere to interpret and teach that story, perhaps along with the state in one of its parks or as another component of Mobile-Tensaw River National Park.

Declarations

Conflict of interest The authors have no conflict of interest to declare that are relevant to the content of this chapter.

Appendix 1: Table of Proposed Parks

See Table 6.1.

Appendix 2: Chronological List of Sources for Proposed Alabama Units

The Office of Planning and Special Studies of the National Park Service in Washington D.C. offers the primary archival data for our study. Unfortunately, some records are poorly maintained, and others have been discarded with personnel changes. However, the planning office has

Table 6.1 List of 72 proposed national park units in Alabama

#	*Name*	*Res.*	*Proposal date(s)*	*Proposal source (s)*	*Reason*	*Current status*
1	Africa town	H	1980, 1983	Mayor of Prichard John H. Smith through Senator Howell Heflin	S	Private
2	Alabama Black Belt	H	2009	NPS; Bill introduced by Senator Debbie Strabenow, D-Michigan	Sig	Private
3	Alabama Forest National Park	N	1935, 1941	Mobile city commissioners through Senator Hugo Black	Sig	Private
4	Athens Battlefield	H	1933	Civil War Sites Advisory Commission	S	Private
5	Bartram Trail	R	1982	Local Bartram Trail groups	F	Mixed private and government
6	Battery Hill Park	H	1939	Unknown	F	Private
7	Birmingham District Heritage Corridor	H	1993	Birmingham Historical Society	S	Mainly private, some local government
8	Black Warrior River	R	1963	John K. Bradley, Southern Timber Management Service, Inc	F	Army Corps of Engineers
9	Bladon Springs	N	1934	Proposed by local citizens through their Congressman John McDuffie	Sig	State park

(continued)

Table 6.1 (continued)

#	*Name*	*Res.*	*Proposal date(s)*	*Proposal source (s)*	*Reason*	*Current status*
10	Blakeley National Historic Site	H	1936, 1941, 1947	John Glennon of Mobile forwarded by Rep. Frank Boykin	Sig	State park
11	Blount Springs Caverns	N	1953	Proposed by W. T. Morgan through Congressman Carl Elliott	Sig	Private
12	Blue Springs	N	1935	Local resident P. F. Whigham	Sig	Private
13	Cahaba River	R	1979	Congress	Sig	Mixed private and government
14	Camp Sibert (or Silbert) including Noccalula Falls	H	1945	Mayor J. H. Meighan of Gadsden proposed through Senator Lister Hill	S	Private
15	Cathedral Caverns	N	1964, 1980	Unknown, but possibly former owner Jacob Gurley after his financial woes	S	State park
16	Chattahoochee Trace National Heritage Corridor	H	2009	Congress	F	Private
17	Chilton County	R	1940	Local farmer wanted to sell 8600 acres	Sig	Private
18	Connors Island	R	1982	Congressman Tom Bevill	Sig	TVA and private

(continued)

Table 6.1 (continued)

#	*Name*	*Res.*	*Proposal date(s)*	*Proposal source (s)*	*Reason*	*Current status*
19	Coosa	A/ H	1933, 1941	Senator Hugo Black was interested in a Coosa Historical Sites National Monument of 10,000 acres and introduced legislation to no avail	Sig	Private
20	Coosa River	R	1989	Congressman Tom Bevill urged several studies in 1988	Sig	Private and state
21	Dauphin Island	R	1936, 1940, 1956, 1979	John Glennon of Mobile forwarded by Congressman Frank Boykin	Sig	Mixed state and private
22	Day's Gap Battlefield	H	1993	Civil War Advisory Commission	Sig	Private
23	Decatur Battlefield	H	1993	Civil War Sites Advisory Commission	Sig	Private
24	De Soto Trail and Mabila	H	1938, 1940	Walter Jones and many historians, archaeolo-gists, and others	F	Private
25	Emuckfaw	H	1958	Unknown	S	Private

(continued)

Table 6.1 (continued)

#	*Name*	*Res.*	*Proposal date(s)*	*Proposal source (s)*	*Reason*	*Current status*
26	Escatawpa River	R	1984	NPS nominated for a wild and scenic river designation	F	Private
27	Flat Rock	H	1930s	Unknown	Sig	Private
28	Fort Gaines	H	1935, 1936, 1399	Herbert C. Brown, Mobile Chamber of Commerce; John F. Glennon of Mobile forwarded by Congressman Frank Boykin	Sig	State historic site
29	Fort McDermott	H	1935, 1941	Nominated with Old Spanish Fort (possibly by Herbert Brown, president of Mobile Chamber of Commerce)	Sig	Privately-owned reserve
30	Fort Mims	H	1936, 1970, 1992	John F. Glennon of Mobile Forwarded by Congressman Frank Boykin; Archaeologists in 1975; Congressman Sonny Callahan in 1992	F	State historic site

(continued)

Table 6.1 (continued)

#	*Name*	*Res.*	*Proposal date(s)*	*Proposal source (s)*	*Reason*	*Current status*
31	Fort Morgan	H	1935, 1939	John F. Glennon of Mobile Forwarded by Congressman Frank Boykin 1936; Mobile Chamber of Commerce	F	State historic site
32	Fort Powell	H	1937	Mobile Chamber of Commerce	F	Destroyed (underwater)
33	Fort Strother	H	1937, 2019	Original unknown; November 2019 Calhoun County Commissioner Eli Henderson sent U.S. Senator Richard Shelby asking for federal protection	S	County
34	Fort Tombecbee	H	1939, 1959	Congressman Peter Jarman	F	State University
35	Fort Toulouse (also called Fort Jackson)	H	1941, 1961	Congressman Kenneth Roberts	Sig	State park
36	Gaineswood	H	1940	Local resident Mrs. W. M. Steele	F	State historic site
37	General Bragg Home	H	1970	Unknown. Proposed for National Register	Sig	Private
38	Helen Keller Home	H	1949	Congressman Robert E. Jones, Jr	S	City historic site

(continued)

Table 6.1 (continued)

#	*Name*	*Res.*	*Proposal date(s)*	*Proposal source (s)*	*Reason*	*Current status*
39	Historic Black Colleges	H	1988	NPS	Sig	State and private colleges
40	Hog Island	R	1982	Congressman Tom Bevill	Sig	TVA, private
41	Holt Reservoir	R	1971	Alabama Historical Commission	Sig	Army Corps of Engineers
42	Jackson Oak	H	1936	John F. Glennon of Mobile forwarded by Congressman Frank Boykin	Sig	City park
43	LaGrange College	H	Before 1957	Local committee and university	Sig	University
44	Lake Bankhead	R	1939	Unknown	Sig	Army Corps of Engineers
45	Lookout Mountain	R	Late 1930s, 1991	NPS	Sig	Private, Alabama Power, state and local parks
46	Mallard Creek	N	1938	Eldridge Wildlife and Conservation Club	Sig	Private
47	Man in Space	H	1984	NASA, Congress, NPS Advisory Board	F	NASA
48	Mobile Bay Eastern Shore	H/R	1934	Congressman John Lister Hill and R. B. Vail, editor of *Baldwin Times*	Sig	Private, state park
49	Mobile Bay Naval Battle	H	1943, 1947, 1993	NPS; Civil War Advisory Committee	F	Mobile Bay

(continued)

Table 6.1 (continued)

#	*Name*	*Res.*	*Proposal date(s)*	*Proposal source (s)*	*Reason*	*Current status*
50	Mobile Bay Shell Mounds (also called Bottle Creek Mounds)	H	1939	Hazel Council, Alabama Federation of Women's Clubs	F	State Historical Site, NHL
51	Mobile-Tensaw River Bottomlands (including Tensaw River and Swamp and Mobile Delta)	N	1960, 1972, 1974, 1979, 1981 until 2016 repeatedly	Scientists, NPS	F	State and private
52	Montevallo Area including Jarmin NP	R	1937	F. H. Hardy wrote a letter to the NPS	Sig	Private and city park
53	Montgomery National Historic Site including First Confederate White House and Liberty Hall	H	1935	Mrs. Minnie-Reese Elmore Richardson	F	Private
54	Moundville	A/H	1941, 1965	Unknown	F	State University
55	Murphy Hill	R	Ca. 1982	Congressman Tom Bevill	Sig	TVA, other federal
56	Muscle Shoals	H	2001	Congressman Robert Cramer	F	Mixed private, state, federal, Natural Heritage Area
57	Newsome Sinks Karst Area	N	1973	Unknown	Sig	Private
58	Noccalula Falls	N	1947	NPS	Sig	City park
59	Oak Mountain Recreation Demonstration Area	R	1936, 1942	Congressman Pete Jarman	Sig	State park
60	Old Cedar Creek Furnace	H	Early 1930s	Unknown	Sig	Private

(continued)

Table 6.1 (continued)

#	*Name*	*Res.*	*Proposal date(s)*	*Proposal source (s)*	*Reason*	*Current status*
61	Phenix City (Ingersoll Hill Battlefield)	H	1935, 1979	State of Georgia passed a resolution calling for a national park in the area in 1935; Phenix City Historical Preservation Society in 1979	Sig	Private
62	Pickwick Lake	R	1935	Alabama Register of Landmarks and Heritage	Sig	TVA and private
63	Pitcher Plant Bogs	N	1980	State government	F	Private (TNC)
64	R. P. Hobson House	H	1970	Unknown	Sig	State historical site
65	Scottsboro, also called Crockett and Jackson NP	H	1966	O. P. Bouldin of Jackson County Forestry Service	F	Mixed local, state, and private lands also including Russell Cave National Monument
66	Selma Battlefield	H	1937, 1993	Congressman Sam Hobbs 1937; Civil War Sites Advisory Commission 1993	F	Private
67	Shelta Cave	N	1984	G, Michael Clark and Henry R. DeSelm letter to NPS	Sig	Private
68	Sloss Furnaces	H	1976	HABS/ HAER (NPS)	Sig	City park

(continued)

Table 6.1 (continued)

#	*Name*	*Res.*	*Proposal date(s)*	*Proposal source (s)*	*Reason*	*Current status*
69	Soldier Creek	R	1984	Congressman Jack Edwards	Sig	Private
70	Spanish Fort	H	1934, 1947, 1989, 1990, 2000	Herbert Brown President of Mobile Chamber of Commerce 1934; John F. Glennon of Mobile 1936 through representative Frank Boykin; Mobile Chamber of Commerce 1940; David Denney 1989	Sig	Private and local government

(continued)

Table 6.1 (continued)

#	*Name*	*Res.*	*Proposal date(s)*	*Proposal source (s)*	*Reason*	*Current status*
71	Talladega or Talladego	H	1931, 1935	G. T. McElder-berry wrote a letter to NPS 1934; in 1940 Women's Chamber of Commerce of Talladega urged consider-ation via Senator Bankhead of a Historical Park of 300,000 acres to commem-orate several battles between 1812 and 1814	F	Primarily private
72	Tennessee-Alabama-Georgia (TAG) Abandoned Railway	R	Early 1980s	Jim Martin, Alabama Commis-sioner of Conservation and Natural Resources, aides of Congressman Tom Bevill and TVA officials	Sig	Private

Res. Indicates resource type: N = natural, R = recreational, H = historic, A = archaeological. Reason refers to why the park was not accepted: S = not suitable, Sig = not significant, F = not feasible compiled lists of proposed areas periodically. The most comprehensive lists from 1934, 1957, 1980, and 1988 can be found in that office's general files, at the National Archives in Suitland, Maryland, or the National Park Service archives in Harpers Ferry, West Virginia. The latter has lists, specific site reports, and other related materials. Other sources include the Congressional Record, the *Federal Register*, archives at individual NPS units, as well as place-specific published articles and National Park Service reports. Some lists and files were able to supply answers to the questions posed above, while others offer only the name of a proposed park.

1. 1934. "Park Planning, Facilities and Lands, Subset F Proposed Areas," NPS Harpers Ferry Center, Proposed Areas Files, Box-Series III, Roger Toll Chart of Inspections, Loose files., *Coosa Historical*
2. 1934. "Proposed Areas," NPS Harpers Ferry Center. L58 Shelf 2, Box containing two reports to Director Horace Albright, ca. 1930 from Arno Cammerer and H.L. Brooks plus a long, detailed list of > 150 proposed areas nationwide. *Coosa Historical, Talladego*
3. 1935-41. NARA, RG 79, Appendix 2, Entry 10, Box 3020, "Proposed National Historic Sites, Alabama, 1935-41." *Fort Tombecbee, Mound Park, Old Spanish Fort, Jackson Oak, Blakeley, Fort Mims, Dauphin Island, DeSoto-Mabila, Fort Gaines, Fort Morgan, Mobile Bay Shell Mound, Coosa Archaeological Site, Battery Hill Park, Flat Rock, Fort Strother, Old Cedar Creek Furnace, Selma Battlefield.*
4. 1935-42. NARA, RG 79, Appendix 2, Entry 10, Box 2922, "Proposed National Parks Alabama, 1935-46." *Spanish Fort, Fort Morgan, Fort Gaines, Battle of Mobile Bay, Alabama Forest, Montevallo, Blue Springs, Lookout Mountain, Pickwick Lake, Mallard Creek, Crockett & Jackson NP (near Scottsboro), Shelby County*
5. 1937, October 23. NARA, Central Classified Files (601), Box 376, Part 5, "Land Acquisition Program Recommended for Historic Sites," *DeSoto-Mabila, Mound Plantation and Sites on the Natchez Trace, Oak Mountain*
6. 1938-47. NARA, RG 79, Appendix 2, Entry 10, Box 2979, Proposed National Monuments, "0-35 Alabama, 1935-47." *Camp Silbert, Chilton County, Dauphin Island, Fort Toulouse, Spanish Fort*

7. 1939. NARA, RG 79, Appendix 2, Entry 10, Box 2939, "Alabama Proposed National Parks, Lake Bankhead, 1939." *Lake Bankhead*
8. 1947, August 29. "Proposed Additions to the National Park System." NPS Harpers Ferry Center, L58, Box 1 and loose files. > 200 proposed areas nationwide. *Alabama Forest, Bladen Springs, Camp Sibert, Coosa hist, Fort Gaines, Fort Morgan, Fort Powell, Mobile Bay Eastern Shore, Mound State, Noccalula Falls, Old Spanish Fort, Phenix City, Talladega*
9. 1949, July 28. "Suggested Area Resumes Region One," Ibid. > 100 proposed units. *Blakeley, Fort McDermott, Fort Toulouse, Gaineswood, Helen Keller Home, Mallard Creek, Scottsboro, Shelby County, Talladega*
10. 1957, January. "Suggested Area Resumes," NPS Harpers Ferry Center, Proposed Area files, including a long and detailed list of > 200 proposed areas. See binder: *Alabama Forest, Bladen Springs, Blakeley, Blount Springs Caverns, Camp Sibert, Coosa Hist Mon, Dauphin Island, Fort Gaines, Fort Morgan, Fort Powell, Fort Toulouse, Gaineswood, Home of Helen Keller, LaGrange College, Mobile Bay, Moundsville, Phenix City, Scottsboro, Shelby County, Talladega*
11. 1958. Edmund Rogers, "Index to Proposals in Congress to Establish National Areas," USDI, NPS. *Fort Toulouse;* For commemorations only *Blakeley, Ft. Mims, Talladega Battle, Emuckfaw war w/ Creeks*
12. 1964. U.S. Department of the Interior, NPS, "Parks for America," (Washington D.C.: GPO). *Fort Toulouse*
13. 1964. "National Park System Plan Studies and Reports, Reject List," NPS Harpers Ferry Center, Proposed Area files, Loose reports 7F72. *Cathedral Caverns. Black Warrior River, Fort Toulouse*
14. 1965, January 18. "National Park System Plan," NPS Harpers Ferry Center, Proposed Area Files, Theodore Swem files. *Moundville historic*
15. 1970. "The Colonial Period in American History, Suggested Sites and Site Types," NPS Harpers Ferry Center, L58, Box 1. *Fort Mims, General Bragg Home, Montgomery National Historic Site, R. P. Hobson House in Greensboro*

16. 1972, June 8. "Authorized and Proposed National Park System Areas," Harpers Ferry Center, Proposed Additions to NPS files, L58. > 150 proposed areas, NPS. *Oak Mtn, Cathedral Caves, Holt Reservoir, Tensaw River Swamp (part of Mobile-Tensaw Bottomlands)*
17. 1974, May 15. Memorandum Acting Chief Scientist, Science Center to Associate Director, Legislation, "Suggested new areas of the NPS-Advisory Board recommendations," Ibid. *Mobile-Tensaw River Bottomlands*
18. 1978, January. "Process for Addition of New Areas to the National Park System," NPS Planning Office, Washington D.C., File-New Area Studies Correspondence 1976-. *Mobile-Tensaw River Bottomlands*
19. 1978, September. "Briefing Statement New Area Study Program," Ibid. *Mobile-Tensaw River Bottomlands*
20. 1978, November 7. Memorandum Acting Chief, Special Programs Division to Manager and Assistant Managers, Denver Service Center, "New Area Study Program and Omnibus Bill Studies Responsibilities," Ibid. *Mobile-Tensaw River Bottomlands*
21. 1980. "Master List of Proposed National Parks," NPS Harpers Ferry Center, L58," Long list of > 200 proposed areas nationwide. *See binder, Africatown, Cathedral Caverns, Dauphin Island, Newsome Sinks Karst Area, Pitcher Plant Bogs*
22. 1980, July 16. Memorandum Director NPS to Assistant Secretary for Fish and Wildlife and Parks, "1980 Section 8 Studies," Ibid. *Mobile-Tensaw River Bottomlands w/data*
23. 1981. "Suggested New Natural Area Units for the National Park System.," NPS Planning Office, Washington D. C., File-New Unit Theme Studies. *Newsome Sinks Karst Area, Mobile River Delta*
24. Early 1980s. Letter Regional Director Baker to NPS Director, "Resource Assessment-Coosa River Area," NPS Planning Office, Washington D. C., File-New Unit Theme Studies, *Cathedral Caverns, Coosa River, Connors Island, Hog Island, Murphy Hill, Tennessee-Alabama-Georgia Railroad*
25. 1982. Richard H. Goodwin and William A. Niering. "Inland Wetlands that Might Warrant Further Study for Possible Inclusion in the National Park System," Ibid. *Mobile River Delta*

26. 1984, May 25. Letter G. Michael Clark and Henry R. DeSelm to Arthur Stewart, Acting Chief National Landmarks Branch, "Natural Landmark Sites." Ibid. *Cathedral Caverns, Shelta Cave*
27. 1984, June. NPS, "Final Wild and Scenic River Study, Escatawpa River," Harpers Ferry Center Library. *Escatawpa River file*
28. 1988, March 21. "New Areas Study Inventory," Long list of > 200 proposed areas nationwide. NPS Harpers Ferry Center, L58. *Dauphin Island, Fort Toulouse, Man in Space, Mobile Tensaw Bottomland*
29. 1989, August 9. Regional Director, SE Region to NPS Director, "Resource Assessment-Coosa River Area," NPS Planning Office, Washington D. C., Proposed Area Files. *Coosa River*
30. 1990. "New Area Studies and Authorizations 1970–1990," NPS Planning Office, Washington D.C., New Area Studies Misc.—Inventory files. *Dauphin Island, Man in Space, Mobile-Tensaw River Bottomlands*
31. 1991, January 16. Memorandum James Ridenour to Assistant Secretary for Fish and Wildlife and Parks, "Follow up with OMB on New Unit Study Criteria," Ibid. *Dauphin Island, Man in Space, Mobile-Tensaw River Bottomlands*
32. 1993, July 10. Civil War Sites Advisory Commission, "Report on the Nation's Civil War Battlefields," Submitted to Congress, July 10, 1993. Financed by the NPS. *Mobile Bay, Blakeley, Fort Morgan, Day's Gap, Athens, Selma, Spanish Fort, Decatur*
33. 1993, August 30. "New Area Study Inventory," NPS Planning Office, Washington D.C., New Area Studies Misc.—Inventory file. *Bartram Trail, Birmingham District Heritage Corridor, Dauphin Island, Fort Toulouse, Mobile-Tensaw River Bottomlands*
34. 1998, February 18. "NPS Special Resource Studies Fiscal Years 1989–1998," Ibid. *Birmingham District Heritage Corridor*
35. 2000. "Special Resource Study Inventory," Ibid. > 475 proposed areas nationwide. (Only names) *Africatown, Bartram trail, Birmingham District Heritage Corridor, Coosa River, Dauphin Island, DeSoto Trail, Fort Toulouse, Historic Black Colleges, Man in Space, Mobile Tensaw River Bottomlands, Old Spanish Fort, Sloss Furnaces*
36. 2007, December. Patrick Gregerson, "Special Resource Studies and Other Authorized Studies-Briefing to the Director NPS," NPS Planning Office, Washington D.C., New Area Studies Misc.—Inventory files. *Muscle Shoals (data on industry and music)*

37. 2016. Ibid., Box 8, "Mounds in Lower Baldwin County." *Chattahoochee Trace National Heritage Corridor, Muscle Shoals*
38. 2016, July. "National Park Service Authorized Studies and Reconnaissance Surveys," NPS Planning Office, Washington D. C., Digital Release, July 2016. *Chattahoochee Trace National Heritage Corridor, Muscle Shoals National Heritage Area*
39. 2023, May 5. NPS Agency (NPS) History, "Proposed Parks," (Agency (NPS) History) *Alabama Black Belt, Bartram Trail, Black Warrior River, Blakeley, Cahaba, Cathedral Caverns, Chattahoochee River, Coosa Historical, Man in Space, Mobile-Tensaw Bottomlands, Montgomery Historical, Newsome Sinks Karst, Soldier Creek, Spanish Fort, Talladego*

NARA = National Archives and Records Administration

*NARA records are indexed in E. E. Hill and R. M Jaussaud, "Inventory of the Records of the National Park Service Record Group 79," NPS, 2007. Author Dilsaver collected most of the Alabama data at NARA in Suitland, Maryland in 2016 and 2019.

References

Alabama Department of Conservation and Natural Resources. 2014. *Alabama state parks: A reflection on 75 years of public service*. https://www.alapark.com

American Battlefield Trust. 2024. *Mobile bay*. https://www.battlefields.org/learn/civil-war/battles/mobile-bay Accessed February 23, 2024.

Carr, Ethan. 2007. *Mission 66: Modernism and the national park dilemma*. Amherst: University of Massachusetts Press.

Conard, Rebecca. 1997. National conference on state parks: Reflections on organizational genealogy. *The George Wright Forum* 14 (4): 28–43.

Dilsaver, Lary M. 2008. Not of national significance: Failed national park proposals in California. *California History* 85: 4–23.

Dilsaver, Lary M. 2016. *America's national park system: The critical documents*. Lanham, MD: Rowman & Littlefield.

Dilsaver, Lary M., and William Wyckoff. 2009. Failed national parks in the last best place. *Montana: the Magazine of Western History* 59: 3-24.

Everhart, William. 1998. Telephone interview by Lary Dilsaver, July 23.

Fenneman, Nevin M. 1916. Physiographic divisions of the United States. *Annals of the Association of American Geographers*. 6 (1): 19–98.

Knight Jr., Vernon James, ed., 2009. *The Search for Mabila: The Decisive Battle between Hernando De Soto and Chief Tascalusa*. Tuscaloosa: University of Alabama Press.

National Park Service. 1990. *Natural history in the national park system and on the national registry of natural landmark*. Natural Resource Report NPS NR NRTR-90 03.

Plumb, Greg E., 2016. A brief history of the mobile-Tensaw river bottomlands national natural landmark. In: Gregory A. Waselkov, C. Fred Andrus, Glenn E. Plumb, *A state of knowledge of the natural, cultural, and economic resources of the greater Mobile-Tensaw River area*. Natural Resource Report NPS/NRSS/BRD/NRR—2016/1243.

Raines, Ben. 2016. Modern cultures of the mobile-Tensaw delta. In: Gregory A. Waselkov, C. Fred Andrus, Glenn E. Plumb, *A state of knowledge of the natural, cultural, and economic resources of the greater Mobile-Tensaw River area*. Resource Report NPS/NRSS/BRD/NRR—2016/1243.

Smith, Brenda. 2003. *National park service U.S. department of the interior criteria for parklands*. National Park Service Web Template Accessed March 8, 2024.

Tabor, Nick. 2023. The forgotten 1980s battle to preserve Africatown. *Smithsonian Magazine*, http://www.smithsonianmag.com/history/the-forgotten-1980s-battle-to-preserve-africatown-180981641. Accessed March 1, 2024.

The Encyclopedia of Alabama: Geography and Environment. 2022. Alabama humanities alliance and auburn university. Geography & Environment—Encyclopedia of Alabama Accessed February 20, 2024.

Tutor, Phillip. 2024. UWA's Dr. Ashley Dumas and students help find Chief Tuskaloosa's territorial province. https://www.uwa.edu/news/the-hunt-for-mabila/. Accessed March 1, 2024.

U. S. Department of the Interior, NPS. 1972a. *Part one of the national park system plan: History*. Washington D. C.: Government Printing Office.

U. S. Department of the Interior, NPS. 1972b. *Part two of the national park system plan: Natural history*. Washington D. C.: Government Printing Office.

University of Alabama, 2024. *Moundville archaeological park*. http://moundville.museums.ua.edu/about/. Accessed March 1, 2024.

Unrau, Harlan D. and G. Frank Williss. 1983. *Administrative history: Expansion of the national park service in the 1930s*. NPS Denver Service Center.

Waselkov, Gregory A., C. Fred Andrus, Glenn E. Plumb. 2016. *A state of knowledge of the natural, cultural, and economic resources of the greater Mobile-Tensaw River area*. Natural Resource Report NPS/NRSS/BRD/NRR—2016/1243.

Wirth, Conrad. 1980. *Parks, politics, and the people*. Norman: University of Oklahoma Press.

CHAPTER 7

Equity and Inclusion in the National Park Service: Historical Challenges, New Possibilities

Yonit Yogev

Introduction and Research Question

In 2016–17 I undertook research entitled, "Diversity, Equity and Inclusion in the National Park Service: Narratives, Counter-narratives and the Importance of Moving Beyond Demographics." The idea for the research germinated from my experience volunteering in 2013 and 2014 at Mount Rainier National Park, near my home in Olympia, Washington. During those summers, I observed that most of the staff and visitors were White. In the ensuing years, I delved deeply into the roots of this observation (validated by research) that approximately 82% of staff and 97% of visitors in the NPS were White (Taylor 2000, 2008, 2014; Taylor et al. 2011).

An extensive literature review (which included a review of mainstream media in addition to academic research) revealed that institutionalized

Y. Yogev (✉)
Independent Scholar, Olympia, WA, USA
e-mail: yonityogev@gmail.com

J. Weber and S. Sultana (eds.), *The Changing Geography of National Parks and Protected Areas*,
https://doi.org/10.1007/978-3-031-74653-6_7

racism[1] was at the root of why communities of color and other underrepresented groups were not visiting public lands in general, and national parks specifically. Initially, that research focused on demographics. If people of color were not visiting national parks and passing along their love of nature to future generations, given the demographic changes occurring in the U.S., the future of public lands was at risk (Floyd 1999).

Over time, research has increasingly examined the reasons people of color were not visiting national parks (and other public lands), using more qualitative research methodologies. Rather than solely examining demographics, research began to illuminate the structural racism and BIPOC visitors' lack of feeling welcomed or safe, which was at the root of the lack of visitation from the perspectives of underrepresented communities (Byrne and Wolch 2009; Finney 2014Shinew et al. 2004). Finally, the research began to specifically name this phenomenon as an issue of environmental *in*justice, since access to these places of beauty and to the inherent healing nature of wild places was constrained (Byrne 2012; Floyd and Johnson 2002; O'Brien and Njambi 2012; Taylor 2008, 2014; Weber and Sultana 2013a, b). Building on these findings, I wanted to approach this research from the community point of view. I also looked at mainstream media sources and found dozens of magazine and newspaper articles, podcasts, films, and books which asked the same questions and posed the same challenges (DuVernay and Averick 2016; Finney 2014, 2016; Grossmann 2010; Nelson 2015, 2016; Peterman 2016a, b, c; Rowland-Shea 2016).

[1] I use the terms 'institutionalized racism,' 'structural racism,' and 'systemic racism' interchangeably. There are subtle differences, but for ease of understanding, I use the following explanation: Institutional racism is "a system of procedures/patterns in education, housing, businesses, employment, professional associations, religion, media, etc., from neighborhood zoning rules separating black and white communities to the development of isolated public housing projects; to federal subsidies for suburban development and neighborhood racial covenants excluding housing access to African Americans; to explicit real estate, insurance, and banking practices, whose effect is to perpetuate and maintain the power, influence and well-being of one group over another. It originates in the operation of established and respected forces in the society; and thus receives far less public condemnation than does individual racism. Researchers have found that much of the wealth and economic inequality gaps existent today among African Americans, Latinos, and Whites can be traced back to segregationist, economic and social practices." (source provided by Jenny Mulholland-Beahrs, personal communication, 2016).

This foundation from the literature led to my research question: What multi-level approaches do people of color and national park staff or partners recommend to successfully increase diversity, equity, and inclusion in the park system, and what specific changes do they think need to occur in order for structural racism to be fully addressed?

In this chapter I review my original research, highlighting the responses and recommendations of the participants. Additionally, I compare the previously noted barriers with those remaining today, highlight positive changes within the NPS and other environmental organizations (EO's), and illuminate other organizational and societal changes which influence this process. Lastly, I share a brief case study of an exceptional program and reflect on what makes it successful.

Methodology/Epistemology

Research that seeks to interrogate and expose social injustice and attempts to right societal wrongs, demands as deeply a democratic process of examination as possible. Qualitative research in general and Participatory Action Research (PAR) specifically bring the human element to the forefront of a research topic. Given that this study examined the complex, nuanced, and strained relationship of BIPOC and other underrepresented communities with public lands, PAR as a critical approach was a perfect fit. I used this methodology and epistemology as the basis for narrative and counter-narrative storytelling from 40 participants in open-ended, unstructured interviews. PAR embodies four essential elements which define and differentiate it from other epistemologies/methodologies (Johnson 2017; Stoudt, Fox and Fine 2012). They are: (1) a community basis; (2) a goal of changing a societal injustice; (3) an inherently democratic, dynamic, and iterative process; and (4) a challenge to the "old" ways of doing research and a question of the assumption of objectivity (Yogev 2017). My research incorporated these elements in the following ways: (1) Underrepresented communities and their experiences were the focus. (2) I began from the assumption that unequal access to natural, healing outdoor spaces is a societal wrong. (3) I used as democratic, dynamic, and iterative a process as possible. (4) I focused on narrative and counter-narrative storytelling to elicit the lived experiences of people of color and park service or partner employees to bring attention to environmental injustice. This inherently challenged old assumptions that quantitative research is the best way to understand human processes and relationships.

I asked participants if they would be willing to be involved in the process from the beginning. Many of them were indeed engaged, from the creation of the research question through to the analysis of the results. I regularly sent updates, but also solicited feedback at every step. Their participation and suggestions made this research endeavor very much a mutual effort. That said, the ongoing introspection required by PAR and constant checking of my positionality made it clear that the final product still carried the privilege I hold as a White, middle-class female who could afford to leave a career and return to graduate school. Thus, my analysis, even with participant input as the central focus, was still ultimately influenced by my own background.

The subject matter also led me to choose Critical Race Theory (CRT) as a conceptual framework for examining the stories and counter-narratives of the participants, and to guide my thinking and analysis. Counter-narratives are a crucial lens into deeper understanding of the centuries of racist and colonialist history embedded in outdoor agencies. Even mainstream media covered, for example, the recent apologies and public reckoning of giants like the Sierra Club and Audubon Society (Brune 2020; Kashwan 2020) with the racism built into their organizations.

Since racism proved to be at the core of this study, I wanted to find a theoretical and conceptual framework which helped situate the research. CRT and PAR are complementary. Both emerged from radical, critical epistemologies deconstructing societal structures that create and perpetuate inequality, governmental policies, and social relations all of which support and prolong inequity and injustice. Both acknowledge and value nuance and context.

CRT first emerged in academia in the 1970s out of legal scholarship. A small group of scholars and activists shared theory, their ideas quickly gained traction, and the group formally convened for the first time in 1989 (Delgado and Stefancic 2012). Its five tenets are as follows: (1) most people of color in the U.S. experience racism daily; (2) gains in civil rights would not have occurred if they had not been in the interests of Whites as well; (3) race is a social construction, and Whites racialize minority populations in different ways at different times depending on the labor needs of the given period; (4) those who experience racism are in the best position to write about it (this spawned the concept of counter-story-telling and counter-narratives); and (5) American history requires a reexamination and retelling from the voices whose histories have not been heard and taught (Delgado and Stefancic 2012).

Because I identify as a White scholar/activist, it was important for me from the beginning to check with participants and receive support and permission to conduct this research. Having received it, it was essential from my standpoint to center the voices of the participants, hear their stories and counter-narratives, and act as an ally to elevate and promulgate their stories. According to Delgado and Stefancic,

"[t]he hope is that well-told stories describing the reality of black and brown lives can help readers to bridge the gap between their worlds and those of others. Engaging stories can help us understand what life is like for others and invite the reader into a new and unfamiliar world" (2012, pp. 47–48).

One of the participants frames it this way:

"...we need to create avenues to let leaders of colors lead, and then follow them and trust their lived experience—as folks who have been...historically marginalized for many, many, many years—to have a perspective of understanding of what needs to be done for justice and equity! POC know what that feels like and have more than a little...experience on ways to come up with solutions..." (Allala, personal communication, 2017).

Results and Discussion

The forty participants were all public lands insiders: national park employees, employees of partner organizations, leaders, and advocates for equitable access to outdoor spaces—many were Black, Indigenous, People of Color (BIPOC). The main themes that emerged from their stories and counter-narratives were absence and marginalization, barriers to access, and solution-oriented recommendations. (Participants who chose to remain anonymous were randomly given numbers, used here when they are quoted.)

Absence and Marginalization

Participants repeatedly spoke of the experience of "not seeing themselves" in national parks, or not seeing "people who look like them." This was apparent not only in the public lands spaces themselves, but also in the media—advertising and documentaries, for example (Finney 2016; multiple participants 2016). The conclusion they drew is that they were not welcome or safe in those spaces. This was also apparent in the stories that parks interpret, but particularly in the stories that were NOT interpreted. One example that a participant shared happened early in their career. Participant #2 began work for an NPS site that manages a cultural estate donated by the descendants of a historic figure. A park guide who oriented the participant took them to a part of the house not typically visited on tours, where the participant noticed an enclosure-like structure. The participant asked about it and the guide informed him that it was used as a sort of holding pen for slaves, probably as punishment, but that the donor family had requested to not include this part of the estate's history on tours. The participant stated emphatically that "withholding stories like that is not helping us move forward in these conversations" (Participant #2, personal communication, 2016).

Other participants shared multiple stories that reflect absence and marginalization. Some shared about the dangers and humiliations their families experienced attempting to travel by car around the U.S. in the 1950s and'60 s. Participant #1 mentioned The Negro Motorist Green Book by Victor Hugo Green, a guide used by Black families to navigate where was safe to travel, which hotels served Blacks and other People of Color, and what places to stay away from, including "sunset towns." Participant #1 stated, "the areas that might cater to African Americans were islands. The ocean itself was hostile to their presence" (Participant #1, personal communication, 2017).

Another participant identified places in the outdoors where Blacks are not welcome, telling the story of a Black family, veteran campers, who were awoken in the middle of the night in a campground to racial epithets and nearby campers chasing them with a shovel (Participant #8, personal communication, 2016) The same participant shared another story of an event they had helped organize with an allied group, bringing together kids and families into nature and exploring "found-in-nature materials." A nearby neighbor took issue with the group and complained they were bringing in invasive species. "And she kept going on and on about 'invasive species'...missing the fact that the kids were having this tactile, beautiful moment with their families in nature...but she kept going on and on about invasive species, and honey, she was not just talking about the plants" (ibid).

Barriers to Access in Employment

There are numerous barriers potential visitors to national parks may face. For this chapter, I focus on the barriers to employment that BIPOC participants shared. This was the second major theme to emerge from participants' stories, usually due to obstacles in hiring practices, recruitment and retention, and funding in public lands agencies. Since many public lands agencies are governmental, significant change to hiring practices is an absurdly encumbered, sloth-like process, often involving multiple government agencies, which further slows any potential change. Since many participants and members of the public see these changes as urgent, some participants suggested solutions such as examining and rewriting position descriptions, some of which can be very biased, using direct hiring authorities, and ensuring hiring managers are aware of these and other options (Beall, personal communication, 2017). Many changes were recommended by participants to the USAjobs.gov site, for example—an antiquated system which requires one have easy access to a computer as well as inside knowledge about how to successfully apply. Recruitment happens in a haphazard way, sometimes through partner organizations, and often stays in-house (current employees inform their contacts, thus perpetuating the same demographic pool). There have been some improvements of late; however, retention remains a problem, in part, due to organizations' lack of a holistic approach to access.

Implicit bias is yet another barrier. If organizations do not address implicit bias in their employees while simultaneously looking at hiring barriers, they may succeed in hiring some employees from underrepresented communities, but those employees, often a sole person of color in an overwhelmingly White workplace, are likely to feel isolated and leave quickly (personal communications, Participant #9, 2017; Participant #7, 2017). It is impossible to overstate how critical this is, and it was emphasized repeatedly by participants. Participant Kieran Slaughter who was an NPS Urban Fellow, a 2-year program designed to connect urban populations with national parks, succinctly stated "the workforce needs to reflect the community it serves and speak the languages that are spoken locally. It creates a comfort zone, and it makes it easier to make cultural connections between visitors and employees" (Slaughter, personal communication, 2016).

Solution-Oriented Recommendations

When recommending solutions, participants made clear that these are not one-off actions, and that they cannot be implemented one at a time. They clearly expressed that these recommendations are interconnected and require a multi-level, multi-faceted approach if outdoor and environmental agencies are to become truly equitable, just organizations.

The solutions fell into the following interconnected categories: (1) Role models and mentors; (2) Targeted environmental education (EE) and outreach (now most often referred to as community engagement); (3) Youth leadership development; (4) Diversity training; (5) Relevancy; and (6) Partnerships and "deep" collaboration.

Mentoring

Many participants expressed the importance of being mentored when they were younger or at the entry level of their careers. They consistently told me how powerful some of these connections and relationships were, especially when they continued over the long term. Moreover, these relationships had the effect not only of modeling mentoring, but also of inspiring in them the desire to eventually become mentors themselves. Additionally, participants emphasized the importance of outdoor agencies and EO's having mentoring programs, and that this is particularly important for retention of employees of color or who are part of other historically marginalized groups.

Targeted Environmental Education (EE) and Outreach

Robert Stanton, the first Black National Park Service director, emphasized courageous leaders at the top who understand the need for long-range, strategic planning "to develop...interest on the part of prospective candidates coming into the work force by engaging young people in conservation programs as far back as junior high or beyond..." (Stanton, personal communication, 2016). Most EE programs focus on youth education. Taking this idea one step further, Latino Outdoors actively engages Latinx families (not just the youth) in outdoor excursions. This holistic approach exemplifies a culturally relevant strategy in which the employees and volunteers come from the same communities and speak the language of the people they are targeting for their programs (Pinon, personal communication, 2017). They found that if the parents are not on board, the youth will not participate. The strong family orientation in Latinx communities dictated a successful strategy for Latino Outdoors and had the added benefit of opening access for more people to national parks and public lands.

Youth Leadership Development

Some EE programs include Youth Leadership Development (YLD). However, many programs lack the staff and funding, or the understanding of its importance, to include this crucial aspect of environmental education. As one passionate environmental educator told me,

> ...NPS conservation staff...often cite exposure to public lands and nature as the key. For example, going on a camping trip, visiting Yosemite, will unlock a world young people will fall in love with and protect. I think this is a deeply false narrative that doesn't...explore the problems with the cultural norms of NPS and how it positions young people...[The youth's] voices and their leadership should be transforming the park service to be reflective of their values, hopes, dreams, and the cultural and natural resources they want to protect (Participant #3, 2016).

They continued,

> They have to care about themselves and their community before they can care about a national park 100 miles [160 kilometers] away. A lot of people miss that message—think they should be doing more restoration work or [be] more focused on climate change. It goes to this idea of relevancy. If it's a true relationship between a park and young people, we should be as open to a conversation that young people want to have, as we expect them to be open to a conversation we want to have. That's what youth development is—giving them a voice.

Another crucial aspect participants talked about is the necessity of long-term relationships with the youth in these programs. Programs which bring kids to a park or green space for a one-time experience, or which change staff frequently, miss out on this all-important part of mentoring and building strength and resilience in the kids. For an EE or YLD program to be most successful, participants emphasized the importance of long-term participation and relationship-building.

Diversity Training (Right Ways and Wrong Ways)

Diversity and inclusion work is a slow, painstaking process. The Oregon-based Center for Diversity and the Environment describes it this way: "…it is an iterative process of awareness building, information gathering, analysis, vision development, planning, and action. There is no magic short cut" (Center for Diversity and the Environment 2024). One intriguing lens into what works and what does not comes, interestingly, from the corporate world. Dobbin and Kalev (2016) and Morse (2016) posit that the environmental world has a lot to learn from the corporate world. Because they pay close attention to the "bottom line," corporations spend more time than government agencies evaluating and analyzing the results of programs and trainings they utilize. Environmental agencies tend to continue using outmoded and ineffective practices and trainings. Dobbin, Kalev, and Morse's research, written up in the Harvard Business Review, as well as the participants' experiences in giving and organizing "diversity trainings" revealed that for them to succeed, the organization must be fully committed to equity and inclusion at all levels. The trainings should occur on a regular basis multiple times a year, should include a variety of learning types, and include formal as well as informal presentations, readings, facilitated discussions, and "regular integration of concepts and conversations into the daily work-life of all the employees at all levels of the organization" (Dobbin and Kalev 2016; Morse 2016).

One of the participants described utilizing that strategy in her environmental organization. Her work group regularly read an article or chapter about equity and justice and then took a few minutes every day to discuss how the material related to their own work and their own implicit biases; these conversations helped employees bring abstract concepts into their daily work. She emphasized that to succeed, there must be support from leadership and from the group, and that leadership has a responsibility to create a safe environment for delicate sharing (Participant #9, personal communication, 2017).

Relevancy

As far back as 2008, the NPS began a long process of planning and introspection resulting from external and internal pressures, in preparation for its centennial celebration in 2016 (NPS.gov n.d). Jon Jarvis, then NPS director, led the agency to examine more closely who was not being served. Under his leadership, the NPS commissioned the Second Century Coalition, composed of NPS staff and outside advisors to work on the Centennial (among many other initiatives during his tenure). Their goal is to view the NPS from the widest possible lens to chart a path forward for the next century (Finney, personal communication, 2017). A key takeaway from years of work was that if the NPS wants to continue to exist, it must continue to be relevant to the people it serves. One critical finding of a 2008–2009 survey was that the NPS needs to be relevant not only to the people who visit the parks, but more importantly, it needs *to become* relevant to those who were *not* visiting.

In 2009, the Citizen Advisory Board added members from the Second Century Commission to continue work on the commission's findings and recommendations. (Former President Trump disbanded the advisory board soon after taking office. It has since been reinstated.) In addition, in 2013, the NPS established the Office of Relevancy, Diversity, and Inclusion (RDI), (recently changed to the Office of Diversity, Equity and Inclusion) with the intention of addressing structural racism strategically throughout the service. According to the Office, "relevancy is achieved when all Americans are able to establish a personal connection to the NPS parks and programs and find meaning and value in the mission of the National Park Service" (https://www.nps.gov/orgs/1244/index.htm, n.d.).

In 1972, the NPS established Golden Gate National Recreation Area, Santa Monica Mountains National Recreation Area, and Gateway National Recreation Area to serve the San Francisco Bay, Los Angeles, and New York metropolitan regions respectively to "bring the parks to the people." During the board and commission's work, they realized that to achieve accessibility and relevancy for underrepresented and potential visitors in the twenty-first century, they would need to scale up these kinds of NPS sites as well as find additional, alternative, "think outside the box" ways to reach underrepresented groups, many of whom had no connection to wildland parks and might not ever. K. Slaughter, the Urban

Fellow mentioned earlier suggested I add a sub-question to my research regarding whether the

> National Park Service [is] telling the stories that are relevant to African Americans...it's not just a relationship with the outdoors that isn't being realized, but stories of national significance that the NPS should be telling. For example, the Black Panther Party was founded 50 years ago this month, it had chapters in almost every state and chapters around the country and world that shaped the Civil Rights movement. Where can people interested in the Panthers see and experience this story? Currently, there is no national recognition...That's one example of how the NPS can be more relevant to people of color, and millennials by making connections that they can identify with and providing an easy entry into the world of the other NPS sites (Slaughter, personal communication, 2017).

Becoming relevant and equitable is a long process for agencies whose antiquated policies and sometimes outright racism are baked into the institutional norms. While positive change is underway, many of the barriers and obstacles still exist.

Partnerships and "Deep" Collaboration

Increasingly, urban NPS sites or partner organizations based in urban centers are working closely with NPS wildland parks. The urban sites create relationships with the youth, often starting with connecting them with nature in their own neighborhoods, adding the all-important element of YLD and environmental justice (EJ), and then slowly bringing them out to the closest wildland parks, where they connect with rangers, learn about those eco-systems, and begin to connect the conservation dots between those more rarified eco-systems and their own neighborhoods. This is a recent change from previous assumptions like that mentioned earlier—that a one-time visit to an exceptionally beautiful place would be enough to change a young person's life.

Charles Beall is the Superintendent of the Seattle National Parks, a group of four sites in the Greater Seattle area. To further the goal of relevancy, he says, the NPS must continue to move in the direction of engaging communities IN those communities. They need to ask the question, "how can the NPS add value to people's lives" (Beall, personal communication, 2017). He brought up a meeting he attended with a community development association to illustrate the point.

"Their questions are not, 'when do the trails open at [the nearby national park,]' or 'what are the hours of your visitor center.' But it's...you know, we just finished this strategic plan about how to improve our community and #1 on the list is we want more opportunities for physical activity—we don't have enough green space here...Anything the park service can do to help there? ...Absolutely!...It's allowing the community to set what's important and meaningful and helping them be successful!" (ibid).

Collaboration is an integral aspect of relevancy. I call the type of collaboration Beall brings forward "deep collaboration" because it goes beyond simply bringing more people to the decision-making table. It incorporates the community in the process from the very beginning. In the past, the NPS connected with communities but would often have a plan laid out and ask the community to rubber stamp the request (Makopondo 2006; Santucci et al. 2014). In the last several years, Beall and others like him in the NPS have taken huge steps toward fostering trust and deep connections with communities, "showing up" as genuine allies. This "deep collaboration," along with the other solutions mentioned above, is necessary for the parks and other EO's to be relevant and equitable, especially important as environmental degradation worsens, and as societal pressures pull youth away from nature.

The barriers and solutions recommended by the participants and validated by ongoing research (Sultana et al. 2023) all point to the need for a major change in government agencies and EOs. Societal norms also need to change and reflect a growing awareness of the need for equitable access to outdoor spaces. One prominent group of researchers, Green 2.0, originally spearheaded by Dorceta Taylor and her colleagues, puts out a yearly report card to check how these agencies and organizations are doing, especially regarding equity and access in leadership roles, as well as boards of directors, employees, and volunteers (https://diversegreen.org, n.d.; Taylor 2014). Despite some improvements, there is still a long way to go before the statistics show full equity for BIPOC in leadership roles in these organizations (diversegreen.org/transparency-cards, n.d.).

As an activist in this field, at times I feel progress is happening at lightning speed. Other times it feels like deep systemic change will never happen. This dance of progress then backlash, more progress, further backlash continues (Rowland-Shea 2016). But it is important to identify where the progress is happening and hope it builds and grows exponentially toward the goal of just and equitable outdoor spaces for all.

Change Is in the Air

Breaking the Barriers

When I began this research in 2015, Outdoor Afro and Latino Outdoors were in their first 2 or 3 years of existence and were nearly the only BIPOC-centered outdoors groups. Some media outlets, and especially journalists of color began to highlight their work, trail breaking as it was. Slowly but surely the coverage increased, until by 2018 or so, there was a groundswell not only of articles and media coverage about equitable access and what it meant, but also of new organizations focused on youth education and developing young BIPOC and LGBTQ leaders in outdoor spaces. Groups advocating for all ages to "get outside" also increased, as did environmental advocacy groups centered on underrepresented communities. Advertising finally got on the bandwagon, and we began to see more BIPOC in advertisements for big outdoor companies such as REI, Patagonia, North Face, and more.

Meanwhile, by 2016, the NPS Centennial year, the RDI office of the NPS[2] had begun to support many parks in their attempts to address implicit bias, in part by offering training (https://www.nps.gov/orgs/1244/index.htm). One such opportunity was Allies for Inclusion, a program which trained park employees in facilitated dialogue, a technique to support safe and deep conversation, to begin to chip away at the biases within the Service at the ground level and help employees communicate and grapple with all kinds of challenges. At the end of that year following Trump's election, he immediately began to undercut and undermine all the work that had begun years before. By the summer of 2017, all such DEI trainings were canceled, funding cut, and the backlash began in earnest. As mentioned earlier, Trump also disbanded the Citizen Advisory Board of the NPS and those of other governmental agencies which had benefited for decades from the public input these boards encouraged.

[2] There has recently been some reorganization and the Office of RDI is now known as the Office of Diversity, Equity and Inclusion.

In some quarters, government employees were forbidden to talk about anything having to do with DEI. Government-funded research ground to a halt. One high-ranking NPS employee quietly told me in 2018 that they were keeping their heads down, continuing what work they could, and holding out until the administration changed.

In May of 2020, the murder of George Floyd forced a reckoning, especially among Whites, who could no longer assume that because we had had a Black president, everything was just fine in terms of progress for Blacks and other People of Color. The need for work on implicit bias once again came to the fore, people spontaneously met with their neighbors, and multicultural training and conversations began happening again with an urgency not seen since the 1960s. Likewise, within the Park Service, conversations began to happen, both spontaneously and with careful planning. And yet, the backlash continued with stunning speed and ferocity.

Backlash and all, however, at least within the park service (and evidently in other EO's), the conversations could no longer be stifled. In addition to attempts at the difficult conversations, so necessary to goals of equitable access, groups began to form in national parks around the country. Some went so far as to establish charters and request official support from their leadership teams. Other parks modeled their groups after the NPS Employee Resource Groups (ERGs), another initiative from the RDI office at the national level.[3] These groups discussed the barriers BIPOC faced in getting into the NPS. They pushed back, they brought it up the ladder, and the RDI office brought speakers from various government or NPS offices to monthly video conversations available to any interested NPS employee. It felt as if this big ship might actually be turning course (multiple personal communications, 2018–2024).

[3] The overview of the guidelines for NPS ERGs states the following: "Employee Resource Groups (ERGs) are employee-led groups with members who are drawn together by a common interest and work towards advancing relevancy, diversity, and inclusion to meet the NPS mission. ERGs benefit employees by increasing connectivity and retention and benefit the organization by expanding recruitment efforts, building more inclusive work environments, and strengthening NPS relevance to the public. These groups also provide developmental opportunities for employees to gain leadership skills. This document will provide information on the benefits of ERGs, expectations, how to form an ERG, how the groups will be held accountable, and how they will be supported" (pdf from nps.gov/rdi, n.d.).

Barriers are certainly still there, but changes are happening, both large and small. Increasing numbers of hiring managers now take advantage of courses on how to change position descriptions to remove biases, as well as how to utilize hiring authorities to diversify their employees (personal communication, 2023). These are still work-arounds rather than root-level changes, at least for USAjobs.gov, but there is a sense that people at the top have heard. Attempts are being made to address another more recent barrier not mentioned earlier, that of steeply rising housing costs, or lack of available housing at national parks. If poorly paid staff cannot afford or find housing in or near national parks, only the privileged few will be able to get "in the door" and begin the climb from seasonal to permanent positions. Many of the changes necessary to remove these obstacles are in the hands of other government agencies, thus requiring a level of inter-agency communication and collaboration that is weighed down by bureaucracy. But with ongoing pressure from within and without and with support from above, these obstacles can be eliminated, and new systems can be built. The process has begun, and hopefully will not once again fall victim to changing administrations, which, as we have seen, can severely impede progress.

Positive Change in the NPS

I have already mentioned some of the positive changes happening in the NPS in the context of removing long-standing barriers. Other changes are also taking place. One of the most receptive departments to the need for change is that of Interpretation. This is the group of rangers who explain, educate, and help visitors connect with what each park offers. Interpretive signs, visitor center exhibits, and displays, as well as programs and guided walks all fall under this department. The participant whose experience I shared earlier in the chapter would today likely find a very different story being interpreted in the estate that had a history of slavery hidden in its attic. NPS employees now openly grapple with and learn how to interpret the sites that have "difficult" stories. Increasing numbers of BIPOC rangers are present in the sites that interpret their stories. Much to my surprise, I counted 73 NPS sites (out of the current 424) which were created, some long ago, most, quite recently, to interpret the previously untold stories of Americans.

Some of the histories and struggles that have been excluded from NPS attention include the Japanese concentration camps, women's suffrage,

Black civil rights and labor struggles, LGBTQ rights, Native burial and other sacred sites, and sites that celebrate or commemorate individuals of color. Many other sites, in particular Civil War memorials or old plantations, tended in the past to avoid discussion about enslavement, and now have begun to interpret those stories from multiple perspectives. For a full list of the sites, see the Appendix.

A particularly horrific, deeply challenging aspect of NPS history is that essentially all the wildland national parks were ancestral native lands violently stolen from the Native Americans who lived there for millennia (Brune 2020; Finney 2014; Makopondo 2006; O'Brien and Njambi 2012; Treuer 2019). The NPS now actively discusses and is negotiating shared administration of some of these sites with Tribal nations. A conference recently convened on this topic, entitled, "Indigenous Co-Stewardship of Public Lands" (George Wright Society e-letter, Feb, 2024).

Other EOs have also realized the grave importance of recognizing and openly discussing their racist pasts. Michael Brune, past Executive Director of the Sierra Club, published a letter in the July 22, 2020 edition of Sierra Club magazine, entitled "Pulling Down Our Monuments" in which he began a series of articles devoted to "a discussion of how the Sierra Club is working to center the voices of people we have historically ignored, so we can begin repairing some of the harms done" (Brune, Sierra Club magazine, July, 2020). In the article, Brune elucidated that Sierra Club's founder and father of modern environmentalism, John Muir, "drew on deeply harmful racist stereotypes [that]...continue to hurt and alienate Indigenous people and people of color who come into contact with the Sierra Club" (ibid). Brune looks back on Sierra Club history as it is interwoven with societal norms and prevailing thought and continues,

> The whiteness and privilege of our early membership fed into a very dangerous idea...that exploring, enjoying, and protecting the outdoors can be separated from human affairs. Such willful ignorance...allows some people to shut their eyes to the reality that the wild places we love are also the ancestral homelands of Native peoples, forced off their lands...before they became national parks. It allows them to overlook, too, the fact that only people insulated from systemic racism and brutality can afford to focus solely on preserving wilderness. Black communities, Indigenous communities, and communities of color continue to endure the traumatic burden

of fighting for their right to a healthy environment while simultaneously fighting for freedom from discrimination and police violence (ibid).

Brune courageously stepped up as the leader of the largest grassroots environmental organization in the United States and admitted to and apologized for the harm done, its lasting effects, and the fact that environmental activists must realize that all equity and social justice struggles are interconnected inextricably with environmental health and the health of the planet.

Months later the Audubon Society published a similar acknowledgment and apology for its racist past, including that of John James Audubon (Kashwan 2020). The Audubon Society, like the Sierra Club, has gone on to make significant changes in its approach, public conversations, and focus. As with the NPS, these organizations continue to struggle with these changes. Declarations from leadership have often come only after years of grassroots pressure and activism. Much of that pressure comes from EJ activists, who were the first to point out the interconnections of environmental and social activism, and who are often those who have historically been excluded from organizations like the Sierra Club and Audubon Society. A lot of work at all levels of an organization, as well as the continued will from within and continued pressure from without (such as from work like Green 2.0's annual "report card" of diversity at organizational leadership and employee levels, must continue for systemic change to take place and take hold. Many other "big" EOs, as well as smaller ones, continue to provide model approaches to creating systemic change.

Case Study: Neighborhood Ecology Corps (NEC), Raleigh, NC

Slowly, environmental, and social justice organizations and agencies learn from their age-old mistakes, and as we have seen, many take seriously their attempts to make amends and create systemic change. I'd like to share one success story because I think this program is a model that should be carefully studied, adapted to individual regions, and scaled up. Mickey Fearn, a past Deputy Director of the NPS and one of the architects of the NEC, has been a mentor and fellow outdoor equity traveler from the time a mutual acquaintance introduced us. A Parks and Recreation expert with 50 + years of experience helping all manner of agencies and organizations learn how they can make systemic change, he continues to passionately inspire his community members and organizations that wisely consult with him.

One of the things that is unique and most powerful about NEC is the number and variety of partners involved. Just about anyone in the community with an interest in equitable access to the outdoors has a place at the table and can contribute from their perspectives and expertise. NEC is a program for urban youth of color in Raleigh, NC. The middle school aged kids first discover nature in their own neighborhoods with EE and community science projects (relevancy). This helps tackle some of the "bio-phobia," Fearn's term for what happens to so many urban youth of color who do not have equitable access to safe outdoor spaces due to long-standing acts of racism and violence that often take place in those same spaces.

After a while these young people are ready to go further afield and get familiar with nature in larger park or wilderness settings. They learn about those spaces and begin to understand how they can bring what they've learned back to their own neighborhoods, with the goal of recognizing and then feeling empowered to change the environmental injustices they see in their own communities. By then, they are hopefully getting comfortable enough (the youth usually participate long-term over about 3 years) to consider things like hiking and camping as a natural part of their recreation inventory. The goal for the graduates is that a well-thought-out curriculum together with their ongoing experiences in local parks, regional, state, and national parks lead them to consider conservation-related careers (Fearn, personal communication 2021).

The collaborations and partnerships that make up the NEC are a remarkable mix of NCSU's College of Natural Resources, local investors and banks, Raleigh Department of Parks and Recreation, the State Department of Parks and Recreation, North Carolina State Parks, and the NPS. No one agency or entity could possibly pull off such a comprehensive program on their own. The long-term deep collaboration, the ongoing relationships—including mentors, educators, and rangers with the youth, are all necessary components of this successful program (https://raleighnc.gov/parks/services/nature-parks-preserves-and-programs/neighborhood-ecology-corps, n.d.).

The follow-up from several years of the program's existence points to success in this goal. When I asked Mickey if the program is working, and if they are tracking the youth, he says,

> Oh yes! The idea is a threshold experience, [then] to sustain that experience, to increase intimacy with the participants, with what they're seeing and doing, that transcends their fear of it, and creates a deep, abiding respect for nature. To understand the relationship of that in dealing with the environmental *in*justice in those communities, because when the conservation movement and the environmental justice movement come together, it's going to be really powerful (Fearn, personal communication, 2021)!

Concluding Thoughts

One of the reasons I continue to publish and present this research is that despite the gains made over the ensuing years, the results are still relevant, applicable. The recommendations of the participants are increasingly implemented in a variety of organizations, and even government agencies, burdened by so much bureaucracy, have begun to make changes. But there is a long road ahead to reach the goal of full equity, full and unquestioned access for ALL to our public lands, and to positions of leadership, or even employment in these agencies and organizations. We have reason to be hopeful, but history shows that people striving for equity in outdoor spaces, as in all basic civil rights, must continue to vigilantly push at the powers that be, and become the powers that be to reach those goals. I remain hopeful as I watch the demographics of my local national park change, as I watch hundreds of new BIPOC-led organizations every year advocating, activating, and becoming the leaders of the present and future environmental agencies and outdoor organizations. Someday soon, the mission of the NPS, to protect and conserve, to allow for the enjoyment of these special places for ALL, will come to fruition.

Declarations

Conflict of interest The author has no conflict of interest to declare that are relevant to the content of this chapter.

Ethical Approval This research was approved by the Human Subjects Review Board of Evergreen State College. Informed consent for participation and future publications and presentations was obtained from all participants. This research does not reflect the National Park Service's position and was carried out completely independently from it.

Appendix: A List of NPS Sites Related to Previously Untold American Stories

African American Civil War Memorial, Washington, DC
African Burial Ground National Historic Landmark, NY
Alibates Flint Quarries National Monument, TX
Amache National Historic Site, CO
Aztec Ruins National Monument, NM
Belmont-Paul Women's Equality National Monument, Washington, DC
Birmingham Civil Rights National Monument, AL
Blackwell School National Historic Site, TX
Booker T. Washington National Monument, VA
Boston African American National Historic Site, MA
Brown vs Board of Education National Historic Park, KS
Camp Nelson National Monument, KY
Cane River Creole National Historic Park, LA
Carl Sandberg Home National Historic Site, NC
Carter G. Woodson Home National Historic Site, Washington, DC
Casa Grande Ruins National Monument, AZ
Cesar E. Chavez National Monument, CA
Chaco Culture National Historic Park, NM
Chamizal National Memorial, TX
Charles Young Buffalo Soldiers National Monument, OH
Clara Barton National Historic Site, MD
Dayton Aviation Heritage National Historic Park, OH
Effigy Mounds National Monument, IA
Emmett Till and Mamie Till-Mobley National Monument, IL, MS
First Ladies National Historic Site, OH
Frederick Douglass Home National Historic Site, Washington, DC
Freedom Riders National Monument, AL
George Washington Carver National Memorial, MO
Gila Cliff Dwellings National Monument, NM
Golden Gate National Recreation Area, CA
Gullah Geechee Cultural Heritage Corridor, NC, SC, GA, FL (Affiliated Areas)
Harriet Tubman National Historic Park, NY
Harriet Tubman Underground Railroad National Historic Park, MD
Honouliuli National Historic Site, HI

Hopewell Culture National Historic Park, OH
Hovenweep National Monument, CO, UT
Hubbell Trading Post National Historic Site, AZ
Inupiat Heritage Center, AK
Knife River Indian Villages National Historic Site, ND
Little Rock Central High School National Historic Site, AR
Lower East Side Tenement Museum National Historic Site, NY (Affiliated Area)
Maggie L. Walker National Historic Site, VA
Maine Acadian Culture, NPS support of ME Acadian Heritage Council site, ME (Affiliated Area)
Manzanar National Historic Site, CA
Mary McLeod Bethune Council House, Washington, DC
Medgar and Myrlie Evers Home National Monument, MI
Minidoka National Historic Site, ID, WA
MLK Jr National Historic Site, GA
MLK Jr Memorial, Washington, DC
Natchez Trace National Scenic Trail, AL, MS, TN
Navajo National Monument, AZ
New Orleans Jazz National Historic Park, LA
Nez Perce National Historic Park, ID, MT, OR, WA
Nicodemus National Historic Site, KS
Ocmulgee Mounds National Historic Park, GA
Pecos National Historic Park, NM
Petroglyph National Monument, NM
Port Chicago Naval Magazine National Memorial, CA
Poverty Point National Monument, LA
Pullman National Historic Park, IL
Rosie the Riveter WWII Home Front National Historic Park, CA
Selma to Montgomery National Historic Trail, AL
Statue of Liberty National Monument, NJ, NY
Stonewall National Monument, NY
Touro Synagogue National Historic Site, RI
Trail of Tears National Historic Trail, AL, AR, GA, IL, KY, MO, NC, OK, TN
Tule Lake National Monument, CA
Tuskegee Airmen National Historic Site, AL
Tuskegee Institute National Historic Site, AL
Wing Luke Museum, NPS Affiliated area, WA

Women's Rights National Historic Park, NY
Wupatki National Monument, AZ.

References

Byrne, Jason. 2012. When green is white: The cultural politics of race, nature and social exclusion in a Los Angeles urban national park. *Geoforum* 43 (3): 595–611. https://doi.org/10.1016/j.geoforum.2011.10.002.

Byrne, Jason, and Jennifer Wolch. 2009. Nature, race, and parks: Past research and future directions for geographic research. *Progress in Human Geography* 33 (6): 743–765. https://doi.org/10.1177/0309132509103156.

Center for Diversity and the Environment. 2024. *Building the foundation: Exploring diversity, equity, and inclusion.* Accessed April 4, 2024. http://cdeinspires.org/building-the-foundation-exploring-diversity-equity-inclusion

Delgado, Richard, and Jean Stefancic. 2012. *Critical race theory: An introduction.* New York: New York University Press.

DuVernay, Ava, and Spencer Averick. 2016. 13th. Netfix Original. Directed by Ava DuVernay. Los Gatos, CA: Netflix, October 7, 2016. www.netflix.com/title/80091741

Finney, Carolyn. 2014. *Black faces, white spaces: Reimagining the relationship of African Americans to the great outdoors*. Chapel Hill: The University of North Carolina Press.

Finney, Carolyn. 2016. It matter who you see in outdoor media. *Outside Magazine*, August 8, 2016. https://www.outsideonline.com/2075586/it-matters-who-you-see-outdoor-media

Floyd, Myron F. 1999. *Race, ethnicity and use of the national park system.* Washington, DC: US Department of the Interior, National Park Service Social Science Program.

Grossmann, James. 2010. Expanding the palette: As America's population become more diverse, will its changes be reflected in park visitors? *National Parks* 84 (3): 1–14.

Johnson, Laura R. 2017. *Community-based qualitative research: Approaches for education and the social sciences.* Thousand Oaks, California: Sage Publications.

Makopondo, Richard O.B.. 2006. Creating racially/ethnically inclusive partnerships in natural resource management and outdoor recreation: The challenges, issues, and strategies. *Journal of Park and Recreation Administration* 24 (1): 7–31.

Nelson, Glenn. 2015. Why are our parks so white? *New York Times*, July 12, 2015. https://www.nytimes.com/2015/07/12/opinion/sunday/diversify-our-national-parks.html

Nelson, Glenn. 2016. Why has the National Park Service Gotten Whiter? *High Country News,* April 8, 2016. https://www.hcn.org/issues/48.14/why-has-the-national-park-service-gotten-whiter

O'Brien, William E., and Wairimu Ngarũiya Njambi. 2012. Marginal voices in "Wild" America: Race, ethnicity, gender, and "nature" in the national parks. *Journal of American Culture* 35 (1): 15–25. https://doi.org/10.1111/j.1542-734X.2011.00794.x.

Peterman, Audrey. 2016a. Needed: A national conversation on the future of our parks. *Huffington Post* (Blog), May 26, 2016. https://www.huffingtonpost.com/audrey-peterman

Peterman, Audrey. 2016b. Fears that 'diverse and qualified are mutually exclusive paralyze change in park service. *Huffington Post* (Blog), August 9, 2016. https://www.huffingtonpost.com/audrey-peterman

Peterman, Audrey. 2016c. Vital questions must be answered on eve of park service centennial. *Huffington Post* (Blog). August 15, 2016. https://www.huffingtonpost.com/audrey-peterman

Rowland-Shea, Jenny. 2016. The rise to power of the congressional anti-parks caucus. Center for American Progress. Accessed April 11, 2016. https://www.americanprogress.org/issues/green/reports/2016/04/11/135044/the-rise-to-power-of-the-congressional-anti-parks-caucus/

Santucci, David C., Myron F. Floyd, Jason N. Bocarro, and Karla A. Henderson. 2014. Visitor services staff perceptions of strategies to encourage diversity at two urban national parks. *Journal of Park and Recreation Administration* 32 (3): 15–28.

Stoudt, Brett G., Madeline Fox, and Michelle Fine. 2012. Contesting privilege with critical participatory action research. *Journal of Social Issues* 68 (1): 178–193.

Taylor, Dorceta E. 2000. Meeting the challenge of wild land recreation management: Demographic shifts and social inequality. *Journal of Leisure Research* 32 (1): 171–179.

Taylor, Dorceta E. 2008. Diversity and the Environment: Myth-making and the status of minorities in the field. *Research in Social Problems and Public Policy* 15: 89–147.

Taylor, Dorceta E. 2014. The state of diversity in environmental organizations: Mainstream NGO's, foundations, and government agencies. Green 2.0 Working Group. Accessed December 20, 2015. https://www.diversegreen.org/wp-content/uploads/2015/10/FullReport_Green2.0_FINAL.pdf

Taylor, Patricia A., Burke D. Grandjean, and James H. Gramann. 2011. *National park service comprehensive survey of the American public, 2008–2009: Racial and ethnic diversity of national park system visitors and non-visitors*. National Resource Report NPS/NRSS/SSD/NRR—2011432. National Park Service, Fort Collins, Colorado.

Treuer, David. 2019. *The heartbeat of wounded knee*. Penguin Audio

Weber, Joe, and Selima Sultana. 2013a. The civil rights movement and the future of the national park system in a racially diverse America. *Tourism Geographies* 15 (3): 444–469. https://doi.org/10.1080/14616688.2012.675515.

Weber, Joe, and Selima Sultana. 2013b. Why do so few minority people visit national parks? Visitation and the accessibility of 'America's best idea.' *Annals of the Association of American Geographers* 103 (3): 437–464. https://doi.org/10.1080/00045608.2012.689240.

Yogev, Y. 2017. *Diversity, equity and inclusion in the national park service: Narratives, counter-narratives and the importance of moving beyond demographics*. Master's thesis The Evergreen State College. https://collections.evergreen.edu/s/repository/item/2587

CHAPTER 8

The Contentious Redesignation of the Gateway Arch National Park

Seth T. Kannarr and Michael A. Urban

Introduction

It can be simply put that 'National Parks' are the best idea we ever had (Stegner 1983). They serve as a mecca of outdoor recreation and appreciation and effuse a reputation of beauty, heritage, and wildlife around the world (Harmon and Putney 2003). Millions of people visit these splendid places every year, and the U.S. 'National Parks' are especially well-known among the American public. As a geographic concept, 'National Parks' serve as centripetal forces that evoke feelings of national pride and bring people together with a sense of unity and shared heritage (Ross-Bryant 2012). These are valuable and cherished places that provide an aura of

S. T. Kannarr (✉)
Department of Geography and Sustainability, University of Tennessee, Knoxville, TN 37996, USA
e-mail: skannarr@vols.utk.edu

M. A. Urban
Department of Geography, University of Missouri, Columbia, MO 65201, USA
e-mail: urbanm@missouri.edu

J. Weber and S. Sultana (eds.), *The Changing Geography of National Parks and Protected Areas*,
https://doi.org/10.1007/978-3-031-74653-6_8

mythos about their history, purpose, and opportunities for all people to partake.

The National Park Service (NPS) in the United States currently manages a total of 429 units, known generally as 'parks' (NPS 2023). Within these 425 NPS units are 19 official and distinct naming designations, ranging from 'National Park' and 'National Monument' to 'National Seashore' and 'National Battlefield'. Although naming designations are theoretically set by NPS guidelines based on the general characteristics of each kind of protected site, the specific designation of a unit may at times be shaped by congressional action instead and can yield contentious results (Condé Nast Traveler 2022).

Recent attention concerning the redesignation of the Jefferson National Expansion Memorial to Gateway Arch National Park in February of 2018 highlights debates about these designation categories (Luong 2018). One side of the argument feels that the designation of 'National Park' is unbefitting of the Gateway Arch due to the lack of 'nature' and its urban setting in downtown St. Louis, Missouri. The other side finds the NPS naming designations to be insignificant and celebrates the redesignation as a boost of attention and tourism to the site.

This speaks to the idea of 'sense of place': what feelings and perceptions that people have regarding a particular location or space (Relph 1976). Critical to 'sense of place' are the words and language we use to describe our feelings and perceptions of the places in discussion (Tuan 1977). When discussing a place, one must use a place name to specify which place they are referring to. Thus, place naming is a central function of sense of place, and it affects how we perceive places. Individuals hold their own perceptions of what values and characteristics 'National Parks' are associated with, and these may or may not match up with the actual characteristics of these places. In this study, I seek to answer the following central research question: how has the redesignation of the Gateway Arch National Park affected its sense of place?

It is worth acknowledging that this study comes from a 'settler' perspective on 'National Parks', as the views and opinions of the common Americans shared here may not be fully representative of all people. Indigenous peoples in particular have a complicated history with U.S. lands, as there is also a movement to remove public access from 'National Parks' and return the land to native tribes (Treuer 2021). These important spaces may mean different things to different communities, and the arguments presented here are not the zenith of thought on this topic. I seek to enlighten understanding of sense of place, protected areas, and place renaming for all people, and intend to do so respectfully for all people.

Criteria for a 'National Park'

The formal criteria utilized by the National Park Service (NPS) and Congress to designate a U.S. 'National Park' are very qualitative and leave a lot of room for interpretation and subjectivity. The NPS criteria used are distinct from other authorities that shed light on what qualifies a national park, in any nation. This begs the question of how to define a 'National Park'. The International Union for the Conservation of Nature (IUCN) is the leading international governing organization for protected areas, nature conservation, and sustainability of nature resources, and has a membership of over 1400 governmental and non-governmental organizations (IUCN 2016). The IUCN identifies a national park as:

> Large natural or near natural areas set aside to protect large-scale ecological processes, along with the complement of species and ecosystems characteristic of the area, which also provide a foundation for environmentally and culturally compatible spiritual, scientific, educational, recreational and visitor opportunities.

This definition stresses the idea of a national park being 'natural', and the idea of ecosystem characteristics and biodiversity as part of a national park. It also notes the scientific opportunities and recreational activities that take place at national parks. Additionally, the IUCN clearly distinguishes national parks as distinct from other 'Protected Area Categories'. A 'National Park' is Category 2, with 'Strict Nature Reserves and Wilderness Areas' as Category 1, and 'National Monuments or Features' as Category 3. The IUCN website details the differences between these categories, and identifies distinguishing features, ongoing issues, and more.

The U.S. National Park Service (NPS) has an array of 429 units that are protected under the Department of the Interior. However, only 63 of these units are designated as 'National Parks'. The NPS details the various naming designations, in which it divides all 19 designations between 'Natural Values' and 'Importance in History' (National Park Service 2016). The designation 'National Park' is nested within the 'Natural Values' category and is described with this language: 'Generally, a national park contains a variety of resources and encompasses large land or water areas to help provide adequate protection of the resources'. In contrast, a 'National Monument' is described as: 'intended to preserve at least one nationally significant resource. It is usually smaller than a national park and lacks a diversity of attractions'. Definitions are provided for other designations as well, ranging from 'Wild and Scenic River' to 'National Historical Park' and 'National Memorial'. These definitions are all vague and subjective, serving as less detailed and clear than the IUCN's definitions and descriptions.

The 'National Park' description provided by the NPS alludes to having a large total area as an important factor, as well as protection and diversity of resources. However, these definitions do not mention the political mechanisms behind their designation. 'National Parks' must be protected and set aside via an act of Congress, while 'National Monuments' are established by executive order via the Antiquities Act of 1906. It is common for a U.S. President to establish a 'National Monument' to protect a location faster than Congress could, and often later 'National Monuments' are then redesignated as 'National Parks' via an act of Congress after a few decades. The two most recent examples of this include 'Pinnacles National Monument' in California (Farr 2013) and 'White Sands National Monument' in New Mexico (Heinrich 2019), redesignated as 'National Parks' in 2013 and 2019 respectively.

It is key to note that the criteria for entry by a protected area into the NPS system is distinct and separate from the criteria for determining the type of naming designation a protected land is given (Vincent 2013). Debates over the designations within the 'National Park System' are not focused on whether a place is intrinsically worthy of being protected under the jurisdiction of the National Park Service, but rather which specific naming designation each NPS unit is assigned and marketed under.

Historical Context of the 'National Park' Designation

The naming designation 'National Park' historically precedes even the establishment of the National Park Service under the Organic Act signed into law by President Woodrow Wilson in 1916 (Dilsaver 1994). Yellowstone was established as the world's first national park in 1872 by Congress under the Yellowstone Act. At the time, the act of setting aside a large tract of public land to be used for conservation and recreation was unprecedented. 'For the benefit and enjoyment of the people' is an aptly used tagline in relation to Yellowstone, a place so special and grandiose that it should be protected forever, under the argument of conservationists like Ferdinand Hayden (Merrill 2003).

The establishment of Yellowstone National Park, to be managed by the Secretary of the Interior, eventually led to many other large significant land areas to be protected by U.S. Congress and designated as 'National Parks', such as Sequoia, Yosemite, Mount Rainier, and Crater Lake. Conversely, the Antiquities Act of 1906 offered a secondary pathway to land area protection by way of an executive order to create 'National Monuments'. Under this act, the President of the United States is granted the authority to set aside public land to be protected as a 'National Monument' via executive order bypassing the lengthy Congressional process for creating a 'National Park'. This was done in response to vandalism of historically and culturally significant Indigenous sites in the United States and has been reserved for land areas much smaller in size and diversity of resources than a 'National Park' would generally entail. Examples include Devils Tower in Wyoming, El Morro in New Mexico, and Montezuma Castle in Arizona, all established as 'National Monuments' under the authority of the Antiquities Act in 1906 by President Theodore Roosevelt.

Over the years, the number of 'National Parks' and 'National Monuments' continued to grow as more areas of land were found worthy of being set aside for protection by the federal government. The Organic Act of 1916 also established the National Park Service which enabled centralized management of this growing system within the Department of the Interior. In 1933, an Executive Order transferred over 50 military parks, cemeteries, battlefield sites, monuments, and memorials from the War Department and Forest Service to be managed by the National Park Service within the Department of the Interior (Dilsaver 1994). This was significant as it greatly expanded the diversity of protected land areas under the Department of Interior and increased the number of unit designations from just 'National Parks' and 'National Monuments', to having 'National Military Parks', 'National Battlefields', and 'National Memorials'. However, it was not until the General Authorities Act of 1970 that the National Park Service would begin to realize its full potential, better organize itself, and create naming designations such as 'National Seashore', 'National Recreation Area', 'National Scenic River', and more to better differentiate the diverse and varied protected areas within it (Dilsaver 1994). Since then, the National Park Service has continued to grow and diversify, and today manages units under 19 distinct naming designations. The complex history of the U.S. National Park system helps to better explain the origin of the 'National Park' designation; and distinguish it from all other naming designations.

Historical Context of the Gateway Arch in St. Louis

The story of the Gateway Arch begins in 1933 when Luther Ely Smith worked with fellow St. Louis city leaders to create a non-profit group entitled the 'Jefferson National Expansion Memorial Association (JNEMA)'. The goal of this association was to create a public and permanent memorial to commemorate westward expansion of the United States, especially the influence of President Thomas Jefferson and the journey of famous explorers Meriwether Lewis and William Clark. The building of a large memorial was also aimed at helping to revive the declining St. Louis riverfront and stimulate the local economy. Discussions about potential funding by the federal government by JNEMA led to an agreement that the future memorial would be managed by the National Park Service.

National economic considerations and dwindling budgets meant that it was not until 1945 that an architectural competition was published to gather potential designs for this public memorial. Eero Saarinen won the competition with his iconic tall arch design. Construction of this massive monument began in 1959 and was not finished until its final dedication in 1967. It took \$13 million to construct the JNEMA, with the \$11 million for the arch itself being funded 75% by the federal government and 25% by the City of Saint Louis (Blahut 2015). The remaining \$2 million for the transportation system inside the arch to bring visitors to the top was funded by the Bi-State Development Agency.

The Jefferson National Expansion Memorial was originally considered an eyesore by residents in St. Louis, but quickly became intertwined with the identity of St. Louis and a famous landmark around the United States (Brown 1984). Despite its rise to prominence, the memorial grounds and infrastructure began to show its age after over four decades in service. In 2010, the Gateway Arch Park Foundation (GAPF) began a capital campaign to revitalize the grounds of the Gateway Arch. This campaign put forward a proposal to connect the Old Capitol building with the

Arch via a greenway as was envisioned in Saarinen's original design and planned to update the Visitor Center and museum underground. This 380-million-dollar renovation concluded in 2018, but not before this campaign by the GAPF pushed for a name change for the site (Fig. 8.1).

The 'Jefferson National Expansion Memorial' name had been long considered unwieldy by visitors and residents alike, and most people refer to the site by its central feature, the Gateway Arch. The GAPF campaign sought a name change to replace 'Jefferson Expansion' with 'Gateway Arch', and to redesignate it from a 'National Memorial' to a 'National Park'. This would highlight and emphasize the work put into the surrounding grounds and refreshed park aesthetic and would better reflect how residents already refer to the structure. It was also suggested that the site's redesignation as a 'National Park' would help attract more

Fig. 8.1 Photo of the Gateway Arch National Park in April 2021. *Source* Authors

tourism and national recognition, as is the case with the recent redesignations of 'Pinnacles National Park' in 2009 and 'Indiana Dunes National Park' in 2019.

During a congressional hearing for the 'Gateway Arch National Park Designation Act', the deputy director of the National Park Service gave a dissenting statement on the matter, noting that the Gateway Arch does not fulfill the NPS's criteria for the 'National Park' designation and would be better suited as a 'National Monument' or 'National Memorial'. Despite this protest, the act proceeded and was signed into law by President Donald Trump on February 22, 2018 (Clay 2018). All advertising, marketing, signage, and information have since been updated to reflect the new official name: 'Gateway Arch National Park'.

Many visitors and outdoor enthusiasts on online forums and social media have questioned this name change, associating the 'National Park' designation with places like Yellowstone and Yosemite. The debate may be strongest within the 'National Park Service Fan Page' community on Facebook, where concerns over the new identity of the Gateway Arch and the relative significance of naming designations within the National Park Service have been raised. This work seeks to explore this debate and describe what values and characteristics are most often associated with making a 'National Park' a 'National Park', or in other words, what sense of place is expected for people at 'National Parks'.

Methodology

Overview

In this project, we sought to identify to what extent the 'sense of place' associated with the Gateway Arch has been affected by its redesignation to a 'National Park'. An individual's sense of a place is fundamentally based on the perceptions and associations that a person holds with a particular location. Thus, it was critical to gather data about how people perceive the general concept of a 'National Park', as well as perceptions of the Gateway Arch National Park specifically. Members of the general public were chosen as subjects in this study, as well as previous visitors to the Gateway Arch to provide a more nuanced perspective shaped by experience. To gather this subjective perception data from these subjects, we decided to assess what values and characteristics people associate with 'National Parks'. We enabled our subjects to describe their perceptions at an easier-to-understand and more accessible level by asking for values and characteristics while allowing us to draw legitimate conclusions about a sense of place regarding these protected areas.

To accomplish these aims, data was collected from an online Qualtrics survey distributed via Facebook, as well as from published Yelp reviews of the Gateway Arch National Park. The Qualtrics survey provides original, primary data from people with an expressed interest in 'National Parks' to use for this research, while the Yelp reviews provide public, secondary data from people who had visited and experienced the 'National Park'. This allowed us to conduct a two-part coding analysis as the principal method for this exploratory case study (Flick 2018). These two datasets together provide a stronger insight into the perceptions that people have regarding the Gateway Arch National Park and allow for a direct comparison of the analytic codes and their respective code counts for analysis. Utilizing the NVivo software package allowed for a straightforward and objective cross-comparison for the coding analysis and to help verify the emergent themes and results. Ultimately, these methods enabled us to translate segmented, written text into usable themes that can more clearly demonstrate how a perceived sense of place has been affected by the redesignation of the Gateway Arch.

Coding analysis can process and condense a large volume of qualitative textual data into simpler, easier to quantify individual codes and themes (Gibbs 2018). These codes are generated to reflect what the intended point and meaning of each response provides, not just highlighting words or phrases appearing the most by volume. We recognize that survey respondents and Yelp reviewers all provide nuanced and diverse perspectives from varying experience levels and familiarity with the National Park Service (NPS) system and that analysis of these texts are subjective by nature (Fig. 8.2).

Survey Creation and Distribution

An online survey was created to assess public perceptions and associations related to the sense of place with the general concept of 'National Parks'. The target audience was adult U.S. residents actively engaged in discussions about their experiences with the National Park Service system. This broad group is more likely to have a strong grasp on the background knowledge and experience necessary to effectively answer survey questions about U.S. 'National Parks' what they value about them and

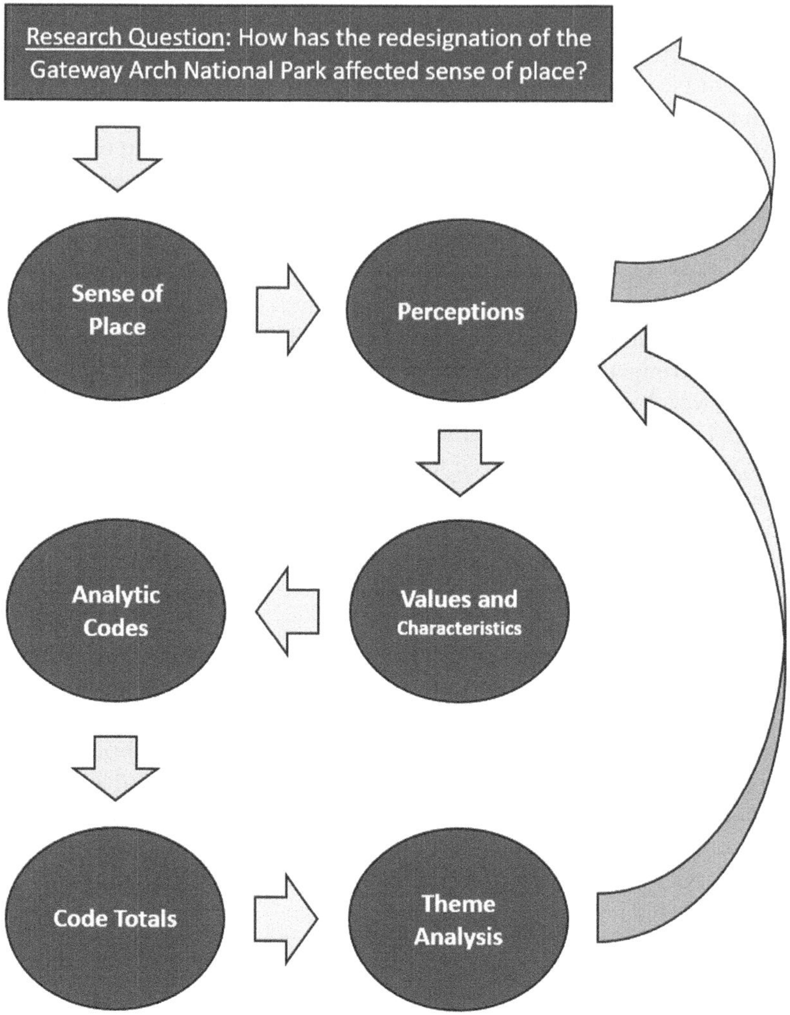

Fig. 8.2 Theoretical framework diagram for this study. *Source* Authors

their experiences with these places. For example, most people outside of this group may not know about the various designations the NPS uses with the 429 units it manages, or how these various unit designations may compare from personal experience. Having this preexisting experience and knowledge will enable survey respondents to have a foundation from which to reflect on their experiences and beliefs as they respond to questions. To find a large volume of these individuals, survey respondents were solicited from the 'National Park Service (Fan Page)' group of the social media platform Facebook, which had over 220,000 members following it as of March 2021 when the Qualtrics survey was launched (National Park Service (Fan Page) | Facebook 2022) (Fig. 8.3).

I utilized Qualtrics survey-building software to construct the survey instrument and solicit online responses. Open-ended questions were carefully designed to avoid potential survey response bias and errors. For example, a multiple-choice question such as 'What do you value about 'National Parks?' would require a list of answer options created by me that would limit the creativity of respondent's answers. Using open-ended questions gives survey respondents the freedom and flexibility to respond with their own thoughts, and not provide a limited set of responses for them to pick between. However, some questions such as 'In which U.S. state do you currently reside?' were drop-down choice questions in which it was practical to provide respondents with the full range of answer options to potentially reduce response processing later, as it can be time-consuming to edit each individual response to ensure that answers like U.S. State names are all spelled correctly, capitalized consistently, and not abbreviated inconsistently.

In seeking to collect data on the perceptions of 'National Parks', we wanted to gather responses regarding the values and characteristics associated with 'National Parks'. We decided to create similar questions that ask survey respondents similar questions in a couple of different ways to help prevent survey respondents from gauging the intent of the survey questions. For example, survey questions 1, 2, and 3 are distinct questions, but all intend to provide qualitative data on the same concept—what perceptions they associate with 'National Parks'. Although not all question alternates were ultimately utilized in this study, it was my intent to provide more questions in advance for potential analysis, as opposed to later regretting not including more questions and being limited to the kinds of responses that could be utilized for analysis in this study. See

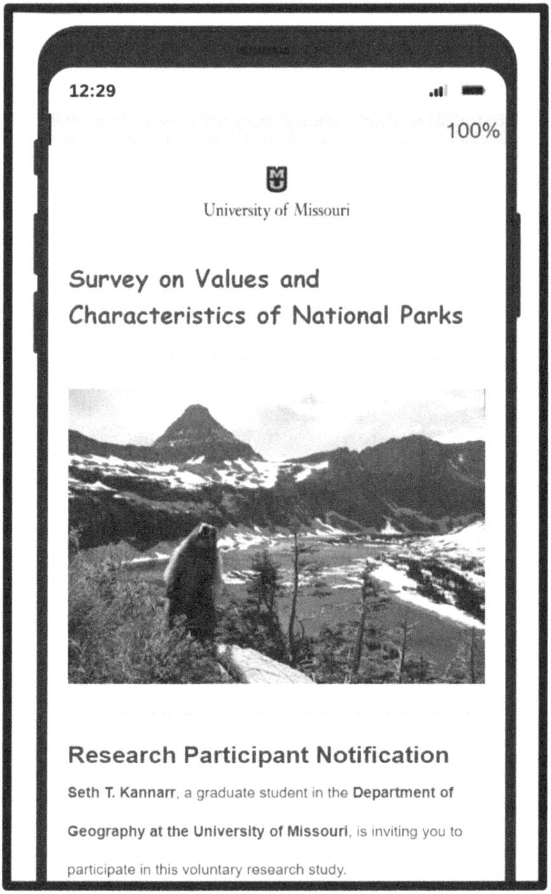

Fig. 8.3 A preview of the mobile version of the Qualtrics survey. *Source* Authors

Table 8.1 for the total list of survey questions utilized, with the relevant questions focused on this research project in bold and italics.

Survey respondents were asked about four specific NPS units under different official NPS naming designations, without identifying them by designation, to create a basis of comparison of values across these different NPS units. This included the Gateway Arch National Park, identified only as the 'Gateway Arch' in direct effort to not let the naming designation

Table 8.1 Total list of survey questions with relevant questions in italics and bold

#	*Question Verbiage*	*Question Type*
0	Do you consent to completing this survey as a participant of minimal to no risk after reading through the MU IRB (Protocol #2054242) text above?	Multiple-choice
1	***What do you value about 'National Parks'?***	***Open-response***
2	Why are 'National Parks' important?	Open-response
3	***What characteristics do you expect a 'National Park' to have?***	***Open-response***
4	What is your favorite 'National Park'?	Drop-down
5	What do you value most about your favorite 'National Park'?	Open-response
6	What are some defining characteristics about your favorite 'National Park'?	Open-response
7	***What characteristics or values do you associate with Yellowstone?***	***Open-response***
8	What characteristics or values do you associate with Mount Rushmore?	Open-response
9	***What characteristics or values do you associate with the Gateway Arch?***	***Open-response***
10	What characteristics or values do you associate with the Statue of Liberty?	Open-response
11	How many of the NPS's 63 officially designated 'National Parks' have you been to before?	Multiple-choice
12	***In which U.S. State do you currently reside?***	***Drop-down***
13	What is your Zip-Code (Five Digits)?	Number fill-in
14	***What is your Age?***	***Number fill-in***
15	***What is your Sex?***	***Multiple-choice***
16	***What is your Ethnicity?***	***Multiple-choice***

influence the answers of survey respondents. Of the four specific NPS unit questions, only Yellowstone was utilized to serve as a 'control' for a 'National Park' and to complement the Yelp reviews data collected for Yellowstone National Park. In addition, survey respondents were asked to provide their U.S. State of residence and Zip-Code to assist in the creation of maps to add context to the results.

The survey was designed with anonymity and privacy in mind, as no personal or identifiable information was collected from any survey respondent. The survey was posted to the 'National Park Service (Fan Page)' group on Facebook via three separate posts, on March 25th, April 6th, and April 12th in 2021. The survey was open for 3 weeks, or 21 days, and 544 complete responses were recorded and used for survey coding and coding analysis. While 625 total responses were collected, 81 responses were started and not completed all the way through by respondents. This incomplete data was securely archived and not included with the complete responses data set for analysis.

Yelp Review Collection

It was most relevant to choose the website most well-known and utilized for place reviews as the depository from which I would pull all reviews of the 'Gateway Arch National Park'. Using the online travel review site Yelp.com, 75 public reviews of the Gateway Arch were gathered since its redesignation as a 'National Park' in February 2018. Reviews were accessed from the site on January 26, 2022. Additionally, the last 75 Yelp reviews for 'Yellowstone National Park' were pulled in the same manner to be used as a direct and parallel comparison to the Gateway Arch. Yellowstone was chosen for three reasons; to serve as a control in this study as it is the original U.S. 'National Park, to provide parallelism in matching with the survey responses from question #7, and because it was selected as the leading favorite 'National Park' by 18.75% of the survey respondents from question #5.

With the Yelp reviews, only the review dates and review texts were collected, with all author and location information omitted. This was done to eliminate potential bias during analysis so that the individual writing the review did not have any influence on the interpretation of the response. Each Yelp review was assigned a number to help identify individual reviews for coding and analysis. The coding process was initiated in the same fashion as they were with the Qualtrics survey responses, utilizing the analytic codes that get at what the intent of each survey response is. These Yelp reviews and the sample batch of manual codes were uploaded into NVivo software for analysis.

Data Coding Process

After collecting survey responses, the next step was to process them and find common threads reflecting the ways in which individuals understand, perceive, and think about 'National Parks'. Analytic codes can be used to succinctly describe and categorize survey responses to better understand trends and themes within the data. It was important to develop a set of analytic codes prior to uploading data into NVivo software for coding analysis, as creating this code index would allow for a consistent approach to coding between both the Qualtrics survey responses and the Yelp reviews. These codes were created inductively from the collected data in an iterative process, starting with a draft and editing to reflect nuances after doing a coding trial with the data.

Here are some examples of how a response would be manually assigned to these codes. A survey response of 'Seeing new animals, hiking, and a beautiful view' is categorized into the codes of 'Biodiversity', 'Recreational', and 'Aesthetic'. Using this approach, a different response with the same intention and meaning, such as 'Unique ecosystems, going camping, and the eye-candy scenery' receives the same codes of 'Biodiversity', 'Recreational', and 'Aesthetic'. However, another example of a response such as 'The beauty is awe-inspiring, it provides me a peaceful quiet on my mountain hikes that I cannot get anywhere else' would receive the codes of 'Aesthetic', 'Emotional', 'Recreational', and 'Experience'. The respondent is identifying the beauty of the protected area, which assigns it 'Aesthetic'. Additionally, the idea of a place being awe-inspiring and providing a peaceful quiet suggest emotional responses and would be coded as 'Emotional'. Lastly, the mention of hikes that the respondent cannot get anywhere else portrays their visit to this protected area as a unique experience, something that is distinct as something to do, thus assigning it 'Recreational' and 'Experience'.

Another element to developing this code index was that of identifying thematic categories that the individual codes could be nested together with. In looking at the range of the fourteen codes utilized, they fit well into three categories: environmental characteristics, intrinsic values, and visitor experience. Environmental characteristics cover what people expect to see in these spaces, or in another way, how they would describe these places. This could range from seeing animals (Biodiversity), to no signs of human infrastructure (Undeveloped), to noting the unique urban architecture present (Manmade). Intrinsic values comprise what may be intangible but valuable or characteristic of these places. This includes 'Historical' or 'Educational' aspects, or even how a visitor may feel from experiencing them (Emotional), or how they believe people should feel about these protected areas generally. This includes mentions of the place's name or naming 'Designation', or even a mention of its value in being protected by the NPS (Protection). The visitor experience thematic category is focused on aspects of a survey respondent or

Yelp reviewer's literal experience that are just as valuable or characteristic of protected areas as that of environmental characteristics or intrinsic values. This includes describing various aspects of visitor 'Accessibility', to what amenities and convenience are available for people (Infrastructure), and especially all the different activities that people want to do—whether indoors or outdoors (Recreational).

Together, these three thematic categories help to provide an outline of how to organize the conclusions to be drawn from the results of the coding analysis in NVivo software. They nest together the individual codes with overarching themes and will allow for more targeted discussion post-analysis with this coding index. See Table 8.2 for the total list of individual codes, the kinds of responses that assign each code, and the thematic categories covering them.

A critical component of this manual coding process includes not discriminating between positive and negative aspects of particular responses. For example, a survey response articulating the lack of COVID-19 safety procedures and supplies like face masks at the Gateway Arch would be coded as 'Accessibility' just the same as a survey response praising the Gateway Arch for excellent minority representation and ease of wheelchair access. The essential idea is that the survey respondent is answering the question from a mindset or approach of accessibility as a value or characteristic of a particular NPS unit or 'National Parks' as a concept. The collection of data around visitor satisfaction or collective attitude regarding user experience is beyond the scope, scale, or interest of this particular study.

For each Qualtrics survey question and Yelp review, the same index of analytic codes was utilized to provide a reference of data to assess how survey responses compare between questions. The same thought process demonstrated before for code assignments was applied to each individual response or case for the following coding analysis. Once this code index was completed, individual documents with the survey responses for each of the four utilized survey questions (#1, #3, #7, #9) and the aggregated Yelp reviews for both the Gateway Arch and Yellowstone were input into the NVivo 12 software to be used for assignment with each survey response and Yelp review.

Each individual survey response and Yelp review, or individual 'case', was carefully evaluated and assigned codes deductively based on the intent and response of the individual. A case may receive at minimum no assigned code, or at maximum assigned each individual code once.

Table 8.2 Thematic categories and codes with descriptions for types of responses

Categories	*Individual Codes*	*Kinds of Responses*
Environmental Characteristics	Aesthetic	Beautiful, Gorgeous, View, Scenic, Spectacular, etc.
	Biodiversity	Wildlife, Ecosystem, Ecology, Plants, etc.
	Geodiversity	Geology, Landforms, Landscape, Features, etc.
	Manmade	Urban Features, Architecture, Manmade, etc.
	Natural	Of/Is 'Nature', Natural, etc.
	Undeveloped	Pristine, Untouched, Wild, etc.
Intrinsic Values	Designation	National Park, 'Not a National Park', etc.
	Educational	Learning, Education, Exhibits, etc.
	Emotional	Patriotic, Peaceful, Religious, Nostalgia, 'Wow', etc.
	Historical	History, Past, Heritage, etc.
	Protected	Set Aside, Managed, Publicly Owned, etc.
Visitor Experiences	Accessibility	Prices, Ease of Access, ADA compliance, etc.
	Experience	Overall trip, Specific programs, Junior Ranger, etc.
	Infrastructure	Museum, Signs, Bathroom, Sidewalks, Trash, etc.
	Recreational	Things to Do, Hiking, Biking, Activities, etc.

The intent was not to gather a frequency count for all the instances an assigned code was invoked, but to see which codes a case response would represent. Due to there being 544 survey responses for each of the four questions utilized, and 75 Yelp reviews for each of the two 'National Parks', 2326 total cases were coded in NVivo software for this research project.

Results

After coding all 2,326 survey responses and Yelp reviews in NVivo software, 4,922 manual assignments of the 15 individual codes were used in this project. This averages 2.12 codes assigned per survey response, with a range of 1 code assigned for a case to nearly all 15 codes assigned for a single survey response or Yelp review (Tables 8.3 and 8.4). NVivo software stored the individual counts for each code per file, and those numbers were transposed into tables displayed in Tables 8.3 and 8.4 to be used for further analysis and discussion outside of the software.

Table 8.3 shows the counts for each code assigned and the thematic categories they fall into for both the Gateway Arch National Park and Yellowstone National Park Yelp reviews. This allows for an easy cross-comparison between the two areas across each code or thematic category. For example, 14.54% of Yelp reviews for Gateway Arch were coded as 'Environmental Characteristics' while 53.85% of the Yellowstone Yelp

Table 8.3 NVivo coding analysis results for the Yelp reviews

National Park		Gateway Arch		Yellowstone	
Total Cases		75	100%	75	100
Total Coding References		227	100%	260	100
Theme	*Envi. Characteristics*	33	14.54%	140	53.85%
	Aesthetics	23	10.13%	34	13.08%
	Biodiversity	0	0%	49	18.85%
Codes	*Geodiversity*	0	0%	45	17.31%
	Manmade	10	3.08%	0	0%
	Natural	0	0%	8	3.08%
	Undeveloped	0	0%	4	1.54%
Theme	Intrinsic Values	90	39.65%	38	14.62%
	Designation	16	7.05%	8	3.08%
Codes	Educational	31	13.66%	4	1.54%
	Emotional	14	6.17%	19	7.31%
	Historical	29	12.78%	3	1.15%
	Protected	0	0%	4	1.54%
Theme	Visitor Experience	104	45.81%	82	31.54%
	Accessibility	22	9.69%	29	11.15%
Codes	Experience	36	15.86%	20	7.69%
	Infrastructure	13	5.73%	8	3.08%
	Recreational	33	14.54%	25	9.62%

Table 8.4 Coding analysis results for the Qualtrics survey

Question		*Q1 Values*		*Q3 Characteristics*		*Q7 Yellowstone*		*Q9 Gateway Arch*	
Total Cases		544	100%	544	100%	544	100%	544	100%
Total Coding References		1442	100%	1394	100%	1103	100%	496	100%
Theme	*Envi. Characteristics*	759	52.6%	593	42.5%	807	73.2%	163	32.9%
	Aesthetics	330	22.9%	200	14.4%	115	10.4%	43	8.7%
	Biodiversity	155	10.7%	105	7.5%	282	25.6%	0	0%
Codes	Geodiversity	103	7.1%	110	7.9%	328	29.7%	0	0%
	Manmade	0	0%	1	0.1%	0	0%	120	24.2%
	Natural	78	5.4%	79	5.7%	34	3.1%	0	0%
	Undeveloped	93	6.5%	98	7.0%	38	3.5%	0	0%
Theme	Intrinsic Values	422	29.3%	395	28.3%	161	14.6%	265	53.4%
	Designation	22	1.5%	4	0.3%	15	1.4%	40	8.1%
	Educational	42	2.9%	165	11.8%	13	1.2%	12	2.4%
Codes	Emotional	95	6.6%	51	3.7%	34	3.1%	38	7.7%
	Historical	76	5.3%	87	6.2%	53	4.8%	172	34.7%
	Protected	187	12.9%	88	6.3%	46	4.1%	3	0.6%
Theme	Visitor Experience	261	18.1%	406	29.1%	135	12.2%	68	13.7%
	Accessibility	61	4.2%	96	6.9%	26	2.4%	16	3.2%
Codes	Experience	125	8.7%	73	5.2%	64	5.8%	17	3.4%
	Infrastructure	21	1.5%	113	8.1%	11	1.0%	22	4.4%
	Recreational	54	3.7%	124	8.9%	34	3.1%	13	2.6%

reviews were coded as such. In the Yelp reviews, 'Environmental Characteristics' are more associated with Yellowstone and 'Intrinsic Values' are more often associated with the Gateway Arch, while 'Visitor Experiences' is only slightly higher for the Gateway Arch. It is worth noting that some codes were not assigned to any responses, being represented with a count of zero. This is the case for the Qualtrics survey responses in Table 8.4 as well.

The same style table was compiled for the Qualtrics survey coding results shown above in Table 8.4 as in Table 8.3 before for the Yelp review coding results. Thus, this table shows the coding and theme totals for all four utilized survey questions. Using the same example as before, 52.64% of respondents from Q1 on Values and 42.54% of respondents from Q3 on Characteristics identified the theme 'Environmental Characteristics' the most for values and characteristics associated with 'National Parks'. This pairs with the same theme accounting for 73.16% of survey respondents for Q7 on Yellowstone and 32.86% of respondents for Q9 on Gateway Arch. As result, more than twice as many survey responses were coded for 'Environmental Characteristics' for Yellowstone over the Gateway Arch, with the values and characteristics percentages lying in between the two 'National Parks'. A similar effect also is shown with the 'Intrinsic Values' theme as well, with the Yellowstone and Gateway Arch roles reversed and values and characteristics in between.

Combined Thematic Code Total Diagrams

After finishing the coding analysis in NVivo software, compiling the coding and thematic category totals, and calculating the significance for each comparison, it became time to transition toward comprehensive analysis and eventual discussion of results. Since this study includes a two-part coding analysis with both primary data from a Qualtrics survey and secondary data from Yelp reviews, it was practical to integrate the datasets together to provide an overall consensus for how the coding totals looked and provide a basis for which to draw discussion and conclusions from.

Figure 8.4 contains the combined totals for each thematic category from the Qualtrics survey responses. This was created by taking the total counts for each theme from Q1 on values and Q3 on characteristics associated with 'National Parks'. Since Q1 and Q3 were developed as analogous questions designed to solicit the collective perceptions and associations with 'National Parks', it was practical to add these totals together to show a baseline of what the collective perceptions and associations are. Under a combined values and characteristics data set, 47.7% of survey responses were coded as 'Environmental Characteristics' between Q1 and Q3. This is the majority of the coding assignments, as 'Intrinsic Values' comes in at 28.8% and 'Visitor Experience' at 23.5%.

Figure 8.5 combines the coding totals from Q7 of the Qualtrics survey responses and from the Yellowstone Yelp reviews to display a combined dataset that represents Yellowstone National Park across this two-part exploratory case study. Results show that 72.1% of survey responses and Yelp reviews coded were within the 'Environmental Characteristics' theme, and only 15.2% of responses under 'Intrinsic Values and 12.7% under 'Visitor Experiences'. These results differ from Fig. 8.4 in that

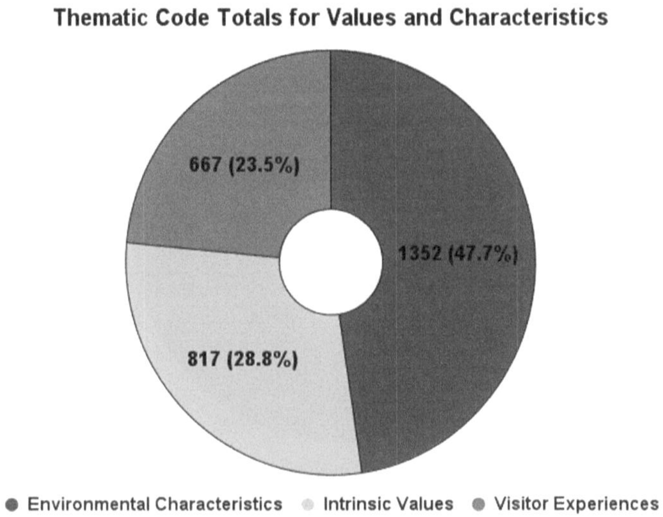

Fig. 8.4 Thematic code totals for values and characteristics. *Source* Authors

'Environmental Characteristics' are at a much higher percentage of the total responses coded, by nearly 25% of the total.

Figure 8.6 is similar to Fig. 8.5 and uses a combined total from both the Yelp reviews and Qualtrics survey responses but was done for the Gateway Arch National Park and not Yellowstone National Park. Results show that 'Intrinsic Values' is in the lead at 49.1% of cases codes, with 'Environmental Characteristics' behind it at 27.1% and 'Visitor Experiences' in last at 23.8%. This differs from Figs. 8.4 and 8.5 in that the Gateway Arch National Park does not have 'Environmental Characteristics' as the leading categorical theme of coding analysis totals. This does align with the results shown in Table 8.3 and 8.4 in that 'Environmental Characteristics' is the leading thematic category for values, characteristics, and Yellowstone National Park, while 'Intrinsic Values' has been the consistent leading theme for the Gateway Arch National Park.

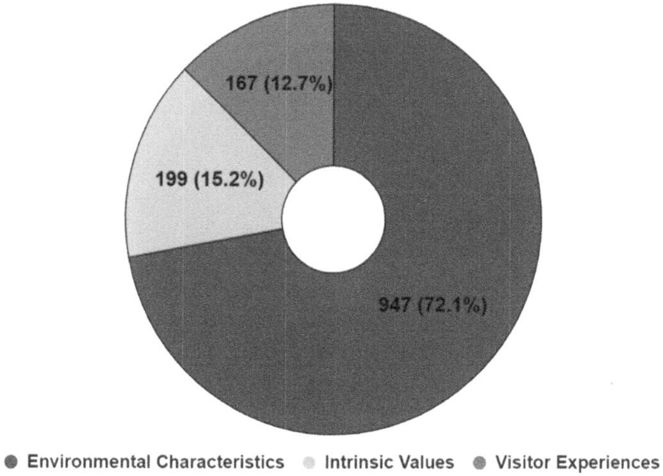

Fig. 8.5 Thematic code totals for Yellowstone National Park. *Source* Authors

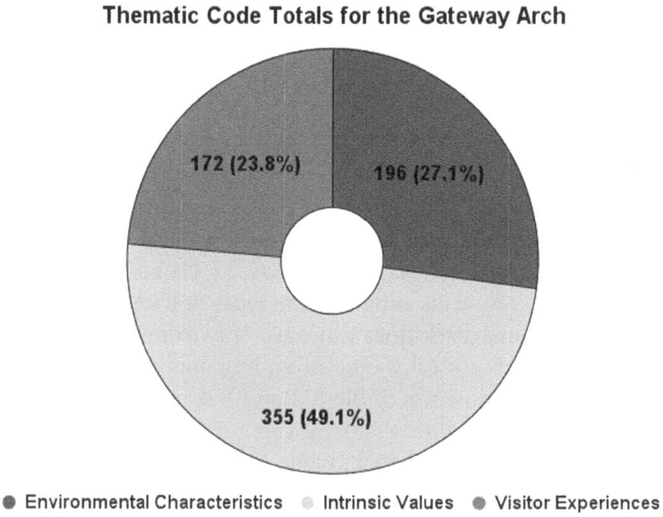

Fig. 8.6 Thematic code totals for Gateway Arch National Park. *Source* Authors

Discussion and Conclusion

The results point to some key takeaways that directly address the central research question of this study. As displayed by the theoretical framework (Fig. 8.2) in the Methodology section, the intention of this study was to extract themes created from the coding analysis of the Qualtrics survey responses and Yelp reviews to gauge the overall perceptions that people have of 'National Parks' in general and the Gateway Arch in particular. The fourteen analytical codes are representative of the range of responses provided, and the three thematic categories serve well as an umbrella for these analytical codes. Some codes such as 'Aesthetics', 'Biodiversity', and 'Geodiversity' stand out as the most common type of response given by study subjects. This does change when looking at the Gateway Arch itself, as 'Aesthetics', 'Historical', and 'Educational' become the most common type of responses. This speaks to the results of the three thematic categories as well, as 'National Parks' have a different proportion of the three themes compared to Yellowstone National Park and Gateway Arch National Park.

Figures 8.4, 8.5, and 8.6 show that the significant majority of survey respondents and Yelp reviewers strongly associate Yellowstone National Park with 'Environmental Characteristics', while it was the theme of 'Intrinsic Values' for the Gateway Arch National Park. However, the collective total of both values and characteristics seemed to lie between the two individual 'National Parks'. Although 'Environmental Characteristics' is the leading thematic category for coding totals of the values and characteristics combined, just like Yellowstone, the difference is exaggerated for Yellowstone as the total of 47.7% of 'Environmental Characteristics' for values and characteristics combined is in between the 72.1% for Yellowstone and the 21.7% for the Gateway Arch. This suggests that although Yellowstone was used as a functional comparison to the Gateway Arch in this study, the combined values and characteristics totals indeed serve as a true control for all 'National Parks'. It is true that the Gateway Arch does not match the thematic categories of the combined values and characteristics as representative of 'National Parks', and although Yellowstone more closely does, it is not a perfect alignment.

The results indicate that people tend to perceive 'National Parks' more broadly as encompassing the variety and range of all U.S. National Parks than just the foundation and standard that Yellowstone set when it was set aside for protection in 1872. Codes such as 'Educational' and 'Historical', which were very common for responses with the Gateway Arch, are more common with 'National Parks' than with just Yellowstone. It would be fascinating to have the ability to go back in time and see if the recent redesignations of 'National Parks' has already influenced the perceptions, or the sense of place, that people have with the 'National Park' designation.

However, the main conclusion is that the values and characteristics most people associate with 'National Parks' do not match those of the Gateway Arch. This is not black and white, as all areas protected under the NPS, regardless of designation, still share basal levels of values and characteristics that come with protected areas in the United States. This is represented in the results as well since the thematic category of 'Visitor Experience' is in third place behind 'Environmental Characteristics' and 'Intrinsic Values' for each comparison. This does not take away from the importance of visitor experiences but instead speaks to this consistent attribute that all these protected areas do share together for all visitors.

What Makes a 'National Park'

If one takes a moment to dissociate these results from this specific case study of the Gateway Arch National Park itself, there is a great opportunity for discussion on the sense of place with 'National Parks', or what makes a 'National Park' itself. The results in Fig. 8.4 show that the combined totals have the three thematic categories in the order of 'Environmental Characteristics', 'Intrinsic Values', and 'Visitor Experience'. Although there are Qualtrics survey responses and Yelp reviews coded with all fourteen analytical codes, some codes were assigned much more often than others. As shown in Tables 8.3 and 8.4, codes such as 'Aesthetics', 'Biodiversity', and 'Geodiversity' may show up the most, all under the 'Environmental Characteristics' theme. Yet the codes of 'Emotional', 'Experience', 'Protection', and 'Accessibility' pop up often as well.

People seem to truly value that these great places are protected and want them to be accessible so that all people can get out there and experience them in all their glory. People of all backgrounds provided emotional responses when considering the values and characteristics of 'National Parks', ranging from nostalgia, religious or spiritual thoughts, patriotism, or even inner peace and calming of the mind. These are protected areas that do an above-and-beyond job of making a significant impact on the people who experience them, as they are truly unique and above-average with characteristics such as 'Aesthetics', 'Biodiversity', and 'Geodiversity'.

I believe that this is central to the sense of place of 'National Parks'. It takes context, experience, or even just a name to turn a point in space into a place. The label of 'National Park' seems to be effective in communicating to people the feelings that these experiences in the natural world can provide on just reputation alone. These feelings paint the context that the experience of going to a 'National Park' can provide. It is this phenomenon that showcases the power of the sense of place with 'National Parks'. It may not matter what the name of the protected area itself is, as all it may take is the naming designation to communicate to someone who has never heard of or been to a place before to know what their experience may be like. These naming designations serve as powerful tools in providing a sense of place that extends beyond the place itself, beyond the values and characteristics that describe the place. It is to this end that thoughtful consideration of which places are deemed worthy of carrying that 'National Park' title is worthwhile, especially with a case study of a place like the Gateway Arch.

Status of the Gateway Arch

Although the redesignation of the Gateway Arch has sent shockwaves and ripples throughout the widespread communities that cherish our 'National Parks', the sky has not fallen, and the world has kept turning. This is not to diminish or support any side in the continual debate over the redesignation of the Gateway Arch, but to remind ourselves that although this is a worthwhile debate, it is not something that has tangible repercussions for the Gateway Arch or the NPS system in the United States. This is all about the naming designations the NPS uses, and the designation currently assigned to the Gateway Arch.

The results of this exploratory case study show us that the values and characteristics associated with 'National Parks' in general, as well as Yellowstone National Park, do not match the values and characteristics of the Gateway Arch National Park. In other words, the Gateway Arch does not align with the values and characteristics that are perceived or expected with 'National Parks' at large. This is especially true for the theme of 'Environmental Characteristics', which was shown to be the majority of associations for 'National Parks' by Qualtrics survey respondents and Yelp reviewers and the minority of associations for the Gateway Arch itself.

The Gateway Arch is much more representative of 'Intrinsic Values' and 'Visitor Experience' over 'Environmental Characteristics'.

Just as a picture can speak a thousand words, direct quotes may paint a different picture than the results of this study can on its own. The following quotes were collected from some of the survey Q9 responses and some of the Yelp reviews for the Gateway Arch. These thoughts on the naming designation of the Gateway Arch were provided freely and without solicitation, as the Qualtrics survey question design was intentional in not referencing naming designations or asking about the naming designations of these protected areas. One survey respondent said: '*A national park label is a stretch, for sure…*' Another wrote '*As a National Park Geek, struggle with the designation here*'. As I continued my coding analysis, they kept popping up. One reads: '*It confuses me that it's a National Park more than a National Monument*'. Another one appears reading: '*Shouldn't be a national park, doesn't define what a national park should be*'. And they kept showing up, with quotes like: '*Why this is a national park and not a national monument baffles me*'. and the final nail in the coffin here: '*Congress made a mistake designating Gateway National Park*'.

These quotes above speak to the intensity to which these people believe in their version of what a 'National Park' is or the sense of place they expect a 'National Park' to impart to visitors. It is key to note though that none of these study subjects directly attacked or criticized the Gateway Arch itself. For them, the issue is the naming redesignation itself, in that it does not align with their expectations for a 'National Park' and that it should not have been ever redesignated as a 'National Park'. The Gateway Arch is a fantastic protected area and offers a great experience for all its visitors. It is okay to disagree with the naming redesignation and still appreciate the Gateway Arch.

Many do believe that all NPS units should be just referred to as 'parks', as it may seem elitist to hold the 'National Park' designation to a different standard or higher value than the other eighteen designations. However, the reality is that the naming designations in the NPS do exist and have different definitions, even if vague and subjective in interpretation. Until that changes, these naming designations matter. Place names matter and play a role in shaping a sense of place. The results and conclusions of this study show that the redesignation of the Gateway Arch from the 'Jefferson National Expansion Memorial' to the 'Gateway Arch National Park' has affected the sense of place that people have with the Gateway

Arch, and that the manifestation of the sense of place into the values and characteristics that people perceive with 'National Parks' does not match with the Gateway Arch. All it would take is another name redesignation to correct this dissonance and settle the debate.

Going forward, the Gateway Arch has a bright future as a cherished protected area managed by the NPS and accessible to all people who wish to enjoy the sense of place the experience offers. If you have not been to the Gateway Arch in St. Louis, it is well worth the experience and memories that the place provides. Regardless of whether it remains a 'National Park' or is 1 day redesignated to more accurately be a 'National Memorial' or 'National Monument', it will continue to offer a window into the westward expansion of the United States in the past and an opportunity to reflect on the present and future of the people of the United States. The Gateway Arch will continue to inspire and offer recreation for people in perpetuity. But, while we're at it, we may as well redesignate it properly, right?

Acknowledgements This book chapter contributed here is an abridged adaptation of my thesis work from the completion of my Master of Arts degree in Geography at the University of Missouri in April 2022 under the advising of Dr. Michael A. Urban, my listed co-author here. This research work was presented many times in 2022, including at the AAG (American Association of Geographers) Virtual Annual Meeting, Southeast Division of the AAG Annual Meeting, and West Lakes Division of the AAG Annual Meeting where I was honored with the 'First Place in Doctoral Oral Presentation Award'.

Declarations

Conflict of interest The authors have no conflict of interest to declare that are relevant to the content of this chapter.

References

Blahut, A. 2015. Review of the gateway arch: A biography. *Buildings and Landscapes: Journal of the Vernacular Architecture Forum* 22 (1): 102–104.

Brown, S. 1984. *Administrative history: Jefferson national expansion memorial national historic site.*

Clay, W. L. 2018. H.R.3058—115th Congress (2017–2018): Gateway arch national park designation act. https://www.congress.gov/bill/115th-congress/house-bill/3058. Last accessed 23 March 2022.

Dilsaver, L. 1994. *America's national park system: The critical documents—edited by Lary M. Dilsaver*. https://www.nps.gov/parkhistory/online_books/anps/anps_1i.htm. Last accessed 23 March 2022.

Farr, S. 2013. Text—H.R.3641—112th Congress (2011–2012): Pinnacles National Park Act. https://www.congress.gov/bill/112th-congress/house-bill/3641/text/pl. Last accessed 23 March 2022.

Flick, U. 2018. *An introduction to qualitative research*. London: Sage Publications.

Gibbs, G.R. 2018. *Analyzing Qualitative Data*. 1 Oliver's Yard, 55 City Road London EC1Y 1SP: Sage Publications Ltd. https://methods.sagepub.com/book/analyzing-qualitative-data-2e. Last accessed 23 March 2022.

Harmon, D., and A. Putney. 2003. *The full value of parks: From economics to the intangible*. London: Rowman & Littlefield Publishers.

Heinrich, M. 2019. Text—S.871—116th Congress (2019–2020): White Sands National Park Establishment Act. https://www.congress.gov/bill/116th-congress/senate-bill/871/text. Last accessed 23 March 2022.

Is National Park Status Always a Good Thing? | Condé Nast Traveler. 2022. https://www.cntraveler.com/story/is-national-park-status-always-a-good-thing. Last accessed 23 March 2022.

IUCN. 2016. Protected Area Categories. *IUCN*. https://www.iucn.org/theme/protected-areas/about/protected-area-categories. Last accessed 23 March 2022.

Luong, Q. 2018. » Gateway Arch National Park: Thoughts on a Name—from QT Luong's Blog. https://www.terragalleria.com/blog/gateway-arch-national-park-thoughts-on-a-name/. Last accessed 23 March 2022.

Merrill, M.D. 2003. Yellowstone and the Great West: Journals, Letters, and Images from the 1871 Hayden Expedition. U of Nebraska Press.

National Park Service. 2016. The National Parks: Index 1916–2016.

National Park Service (Fan Page) | Facebook. https://www.facebook.com/groups/44423080831. Last accessed 23 March 2022.

National Park System (U.S. National Park Service). https://www.nps.gov/aboutus/national-park-system.htm. Last accessed 9 July 2023.

Relph, E. 1976. *Place and placelessness*, 1st ed. London: Pion.

Ross-Bryant, L. 2012. *Pilgrimage to the National Parks: Religion and Nature in the United States*. New York: Routledge.

Stegner, W. 1983. *National park service history: Famous quotes concerning the national parks*. https://www.nps.gov/parkhistory/hisnps/npsthinking/famousquotes.htm. Last accessed 23 March 2022.

Treuer, D. 2021. Return the national parks to the tribes. *The Atlantic*. https://www.theatlantic.com/magazine/archive/2021/05/return-the-national-parks-to-the-tribes/618395/. Last accessed 18 April 2022.

Tuan, Y.F. 1977. *Space and place: The perspective of experience*. London: University of Minnesota.
Vincent, C.H. 2013. *National park system: Establishing new units*. https://crsreports.congress.gov/product/pdf/RS/RS20158

CHAPTER 9

Canyon de Chelly National Monument and the Vanishing Navajo Meridian

Joe Weber

Introduction

Arizona's Canyon de Chelly is one of several connected canyons on the west side of the Chuska Mountains in the northeast corner of the state. The two largest canyons, de Chelly[1] and del Muerto, converge near the mouth of the canyon system. Monument and Black Rock Canyons are two major tributaries, and many smaller ones also feed into the system. The canyons are shadowed by sandstone cliffs towering as much as 1000 feet (305 m) above the canyon floor. The slender spire of Spider Rock stands almost 800 feet (244 m) high at the confluence of de Chelly and Monument Canyons (Fig. 9.1). Within these sheer cliffs, many overhangs or caves have been carved by water erosion.

The canyons are also noteworthy for their human history. Here the Ancestral Pueblo (or Anasazi) people built many farming villages or

J. Weber (✉)
Department of Geography, University of Alabama, Tuscaloosa, AL 35401, USA
e-mail: jweber2@ua.edu

J. Weber and S. Sultana (eds.), *The Changing Geography of National Parks and Protected Areas*,
https://doi.org/10.1007/978-3-031-74653-6_9

Fig. 9.1 Photo of spider rock in Canyon de Chelly. Photo by Jim Weber

pueblos between 900 and 1350, many still standing as ruins (Rohn and Ferguson 2006). Since the 1700s the canyon system has been occupied by the Navajo, or Diné, for whom it provided not only a suitable place for agriculture as well as a defensive refuge, as during incursions by the Spanish Empire in 1805 and the U.S. army in 1864. Its combination of scenery and history attracted many Anglo explorers, and in 1931 this interest resulted in it being proclaimed a National Monument by President Hoover, though an unusual one in that the land remained the property of the Navajo Indian Tribe despite being managed by the National Park Service (NPS). It remains a spectacular, though underappreciated, unit of the nation's park system.

This paper discusses the creation and alteration of boundaries for Canyon de Chelly, focusing on the question of how outside agencies used these boundaries to impose legibility on this landscape, and how each attempt influenced later boundaries. It draws on General Land Office plats, National Park Service maps and land descriptions, and other cartographic sources to tell an interlinked story about a National Monument, an Indian Reservation, and the Public Land Survey System and their relation to this special place.

Background

Each of the 429 units of the national park system has legally defined boundaries which delineate the land and resources inside the park, protected by the National Park Service (NPS) from the world outside. These boundaries can be controversial, may be changed, and are also tools to achieve management goals. A crucial part of constructing boundaries is the concept of legibility, first discussed by Lynch (1960) as a way of making sense of cities. He described it as 'the ease with which its parts can be recognized and can be organized into a coherent pattern' (pp. 2–3). This will vary among groups and cultures and is subject to constant revision. More recently Scott (1998) has used legibility as the process by which governments impose a simplified order, ignoring local customs and practices, in order to more easily identify, classify and organize, measure and manage people and resources. Classifying vegetation, surveying land into a grid, or grouping people into distinct tribes will have little relevance to the lives of the people in that area but will give managers greater control over them. This is related to Sack's (1986) discussion of territoriality, in which a group puts forth a claim to a particular location and what may be done there by whom. Such a claim requires some act of classification of what is and is not included, some way of demonstrating this (an oral legend, a fence, or a map, for example), and efforts to control the area. Territoriality could apply to Native American groups or government agencies, though with considerable differences as to the degree of control exerted and how this would be communicated.

The greater the intervention of the government, the greater the legibility required. Legibility can lead to planning disasters when attempts at imposing order are accompanied by a modernist ideology that ignores local practice and knowledge, an authoritarian state intent on putting its plans into action in the face of a society that lacks the resources to resist them. While Scott applied this approach to understanding spectacular planning disasters such as forced rural collectivism, it is relevant to the case of Canyon de Chelly. The canyons have been subjected to a series of boundaries based on vastly different geographic concepts, several of which are clear expressions of a goal to make a strange and often baffling landscape legible to government workers far removed from the canyon. In such landscapes, there will be competing perspectives that justify different priorities and activities (Jenkins 2018). Legibility is central to many of the activities of the NPS to interpret a landscape for the public (Feldman 2011). At Apostle Islands National Lakeshore, the landscape was first made legible by state regulation of economic activities designed to bring about greater efficiency; later the NPS created a master plan that assigned uses and interpretive themes within the park. Designating an area as a natural environment or wilderness was not simply a recognition of its condition but reflected a strategy of removing cultural features from such areas as well as providing information to the public about what they can expect to see and experience in different areas. The public view is in some ways an illusion created by the NPS designed to focus attention on certain themes or issues and not others, and which does not recognize the long history of land use and resource extraction in what appears to be wilderness.

The People

The Navajo or Diné are the largest federally recognized Native American tribe in the United States. The Navajo traditionally defined their homeland as that territory lying between four sacred peaks, commonly identified with the San Francisco peaks in the west, Mount Taylor in the south, Blanca Peak to the east, and Hesperus Mountain in the north (McPherson 1992). The Navajo worldview and religion are strongly tied to the landscape, with physical features having a spirit. The landscape is not just inanimate scenery but a living thing. Sacred mountains should not be climbed except after proper precautions have been taken; their sacredness may be destroyed if they are logged, developed for recreation, or have roads built on them. Shiprock was sacred until 1939 when a Sierra Club group climbed it and destroyed its power (McPherson 1992).

Although important differences exist between de Chelly and del Muerto in terms of their potential for farming, today both canyons are seasonally occupied by perhaps several hundred Navajo who have their primary residence above the rim (Grant 1978; Andrews 1991). Many families have ownership claims in a canyon that can be traced back well over a hundred years. From a rim overlook the canyon floor may seem inaccessible, but there are dozens of trails that allow for foot or horseback travel between the canyon bottom and rim (Jett 1994, 2001). They are part of an interconnected Navajo cultural landscape (Jett 1978, 1993; Iverson 2002). Though not shown on any map, a vast number of Navajo place names exist within the canyon system (Jett 2001). These differ substantially from many common national park names in that the Navajo do not name places after people or events (Linford 2000). The canyons are also a part of many ceremonies, and Spider Rock and White House Ruins are perhaps the most sacred places within the canyons (Linford 2000). The Navajo are the second people to live in the canyons, and the Anasazi ruins found here and throughout the Four Corners area are important to the Navajo (McPherson 1992). In Diné tales the Anasazi were often destroyed by their pride and arrogance or by not following the correct spiritual traditions; their ruins, graves, and artifacts contain evil spirits and are best avoided. The Navajo traditionally therefore avoided Anasazi ruins, though attitudes have slowly changed. While by the 1990s Navajo workers made up much of the labor force at Navajo National Monument, which is also located within the Navajo Reservation and preserves spectacular Anasazi cliff ruins, they were only recently becoming tourists (Rothman 1991).

The Reservation

The Navajo Indian Reservation was created in 1868 by a peace treaty signed between representatives of the tribe and the U.S. Government (Kappler 1904; Kessell 1981; Iverson 2002). At that time most of the tribe was in confinement within the Bosque Redondo reservation near Fort Sumner, New Mexico, following an 1864 scorched-earth campaign led by famous explorer Kit Carson. His troops drove the Navajo out of Canyon de Chelly and cut down the peach trees they had carefully tended in the canyon bottoms, forever shattering the image of the canyons as a secure defensive refuge.

The tribe was given a choice between a reservation in their homeland or one in Indian Territory (now Oklahoma), and they chose the former. Their new reservation boundaries were defined as: 'bounded on the north by the 37th degree of north latitude, south by an east and west line passing through the site of old Fort Defiance [Arizona], in Cañon Bonito, east by the parallel of longitude which, if prolonged south, would pass through old Fort Lyon [now Fort Wingate, New Mexico], or the

Ojo-de-oso, Bear Spring, and west by a parallel of longitude about 109 degree 30' west of Greenwich, provided it embraces the outlet of the Cañon-de-Chilly, which cañon is to be all included in this reservation' (Kappler 1904, 1016). The reservation's northern boundary would be the northern boundary of the Arizona and New Mexico Territories. Though Canyon de Chelly was mentioned along the western boundary it was apparently intended that Chinle Valley be the western edge of the reservation (Kessell 1981); presumably for this reason the western boundary was surveyed farther west at 109 degrees 37 min 36 s west longitude (Fig. 9.2). Intentions and results were not always in geographic alignment when it came to nineteenth-century reservation boundaries (Hart 2008). The first expansion of the reservation, in 1878, was intended to bring the entirety of Chinle Valley into the reservation.

The canyon was one of the few areas on the reservation suitable for agriculture and Navajo men who were heads of households wishing to farm were allowed to claim up to 160 acres (65 hectares) of land in the reservation, while other men were allowed to claim up to 80 acres (32 hectares). This provision long preceded the 1887 Dawes Act, which sought to eliminate reservations by turning them into private property. These actions required a land survey so that individual allotments could be passed into private ownership; for that, the Public Land Survey System [PLSS] was to be applied.

The Survey

The PLSS was adopted by the United States in 1785 as a means of quickly surveying the new country's western frontier so that land could be sold to farmers (White 1983; Hubbard 2009). An invention of Thomas Jefferson, the system was based on the surveying of square townships six miles (9.7 kilometers) to a side. These could be subdivided into one-square mile (259 hectare) sections and later into a variety of smaller square or rectangular subdivisions down to 40 acres (16 hectares) for sale or homesteading (Webster and Leib 2011). As new lands were acquired by the United States the survey was extended to them, often beginning with the establishment of a new baseline and meridian from which townships would be laid out. The complete survey of the western frontier (excluding Texas and Hawaii) eventually involved many separate meridian and baselines, the last being established in Alaska in the 1950s (Weber 2023).

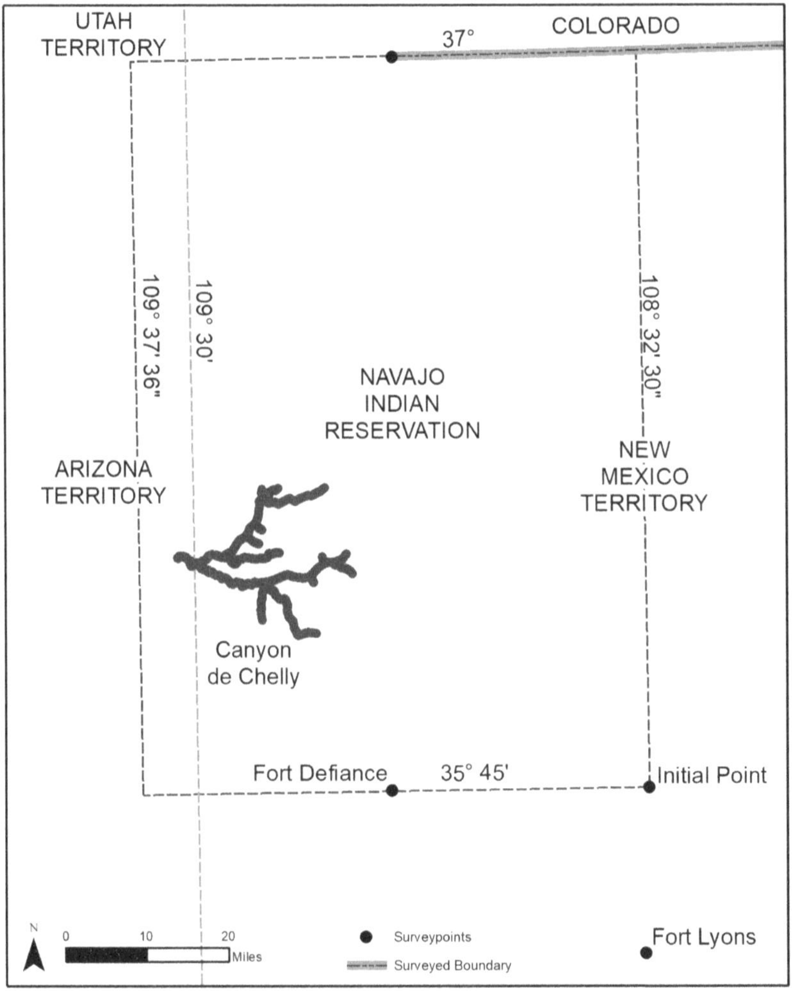

Fig. 9.2 Map of Navajo reservation, 1868. Map by Author

Indian Reservations had an awkward relationship with the survey; they were initially skipped over by surveyors as these lands were closed to white settlers looking to buy or claim land, but then later surveyed when policy shifted toward encouraging Native Americans to become farmers on individually owned land (Pearce 2004; Winslow 2013). When this was done in western states a new meridian and baseline would sometimes be created just for a reservation, in effect creating a dual system of surveys for white and native farmers (Weber 2021). This option was selected for the new Navajo Reservation.

In 1869 the General Land Office, which conducted surveys, hired Ehud Darling (1832–1912) to survey the reservation. A surveyor from Minnesota, Darling had completed a survey of the New Mexico-Colorado boundary the previous year and was familiar with the country (Roeder 2009). Since the reservation was remote from surveyed areas, he established the Navajo Meridian and Baseline, at an initial point at the southeast corner of the reservation, which he determined astronomically. From here he and his crews surveyed the boundaries of the reservation, including the northern border along the Arizona-Utah boundary, as this boundary had not yet been surveyed. Two standard parallels and a guide meridian were established, following standard procedure; these were correction lines used to account for the curvature of the Earth (Hubbard 2009). Then crews surveyed the exteriors of 78 complete townships, all to the north of the baseline and to the west of the meridian. His final map (Fig. 9.3) shows that surveyed townships make up only about half the area of the Navajo Reservation; following common procedure Darling skipped mountainous areas or other lands unlikely to support agriculture and provided few details about the land he traversed. Ironically, Darling did not survey Canyon de Chelly, the only area in the reservation where agriculture had been regularly practiced. After completing his surveys, he secured a contract to survey townships for the Indian Meridian in Indian Territory (now Oklahoma) in 1870 before returning east and into obscurity.

The National Monument

America's national park system began in the second half of the nineteenth century when Congress set aside land in the public domain to preserve scenery (Weber and Sultana 2024). In 1864 (the year the Navajo were forced out of Canyon de Chelly) the granite valley of Yosemite was set aside by Congress for its scenery, though under the supervision of the state of California (Runte 2010; Mackintosh, McDonnell, and Sprinkle 2018). By 1906 ten national parks had been created, almost all in the

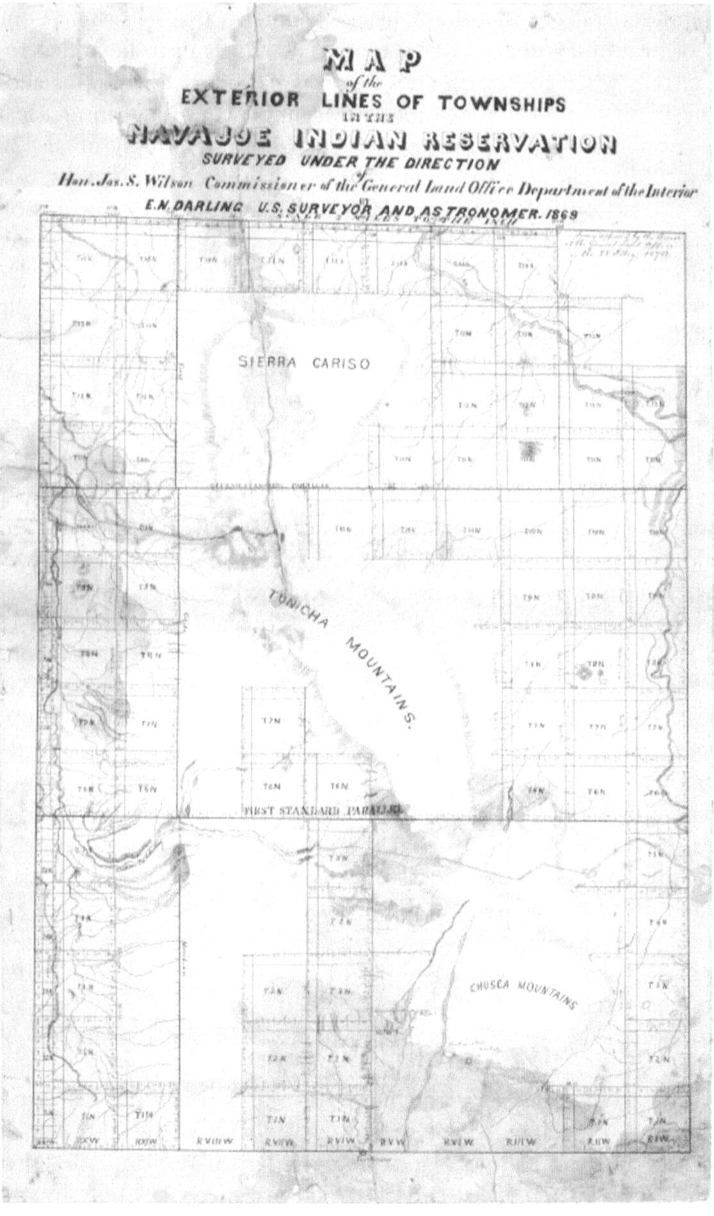

Fig. 9.3 Map of 1869 Navajo baseline survey. *Source* Bureau of Land Management (2020)

west and generally in mountainous environments that fit eastern conceptions of beautiful scenery. A new type of place began to be added to the park system when the Antiquities Act was passed in 1906. This allowed the president to proclaim a National Monument (NM) around a feature of scientific interest (Rothman 1994a, 1994b; Harmon et al. 2006; Sellars 2007). The Act had been motivated by the discovery by eastern archaeologists of the spectacular prehistoric Native American ruins of the Southwest, including Mesa Verde in southwest Colorado (which became a National Park) as well as Arizona's Casa Grande Ruins, though Devils Tower, a geologic oddity in Wyoming, was the first to be created. Canyon de Chelly was among the Native American archaeological sites protected; though it had been investigated by those interested in its prehistory since the 1870s Canyon de Chelly National Monument was not created until 1931, after several years of discussion between the National Park Service, the Navajo Tribe, and Bureau of Indian Affairs, which oversees tribes and reservations (Sullivan 1947; Brugge and Wilson 1976). The monument is unusual in that the land remains the property of the Navajo tribe, unlike Navajo NM and Hubbell Trading Post National Historic Site, located elsewhere on the Navajo reservation but owned by the NPS.

Archaeology was the primary purpose of the park, with the ruins as the main attraction (Brugge and Wilson 1976). The mouth of the canyon was the natural place to locate the monument headquarters as well as visitor services. The ruins were located on the canyon floor and roads up the canyons were planned; these were soon dropped because of the threat of flooding and objections by canyon residents. Instead, tourists would be kept above the rim where they would view the ruins at overlooks. Roads were to be provided from the headquarters along the north and south rims to various overlooks. The south rim road was built first and included an overlook for the White House Ruin below. Visitors at the overlook may hike down the ruin on the White House Trail. This was built in 1934 and was considered a major attraction in the park. It was, and remains, the only place where unguided visitors can reach the canyon floor (though it was temporarily closed in 2020 and remains closed at present).

At archaeological parks, the NPS has often paid little attention to living tribes in the area (Keller and Turek 1998). Canyon de Chelly is no exception, and the Navajo Nation regretted their decision to allow the monument and acted at least twice to remove the NPS, though without success. Their grievances were based on an NPS management style that ran roughshod over the desires of residents and traditional users of the canyons. While Navajo generally supported the monument because of promises the NPS would build good roads and the opportunity to make money off tourists, they were not happy with those tourists (Brugge and Wilson 1976). One example was the first ascent of Spider Rock by white climbers in 1956, an event the tribe was opposed to as the Rock is sacred.

The First Boundary

Every new national park unit requires the specification and surveying of legal boundaries. In the western part of the country national park units are usually based on the PLSS, giving them boundaries made up of straight lines running in cardinal directions (Weber and Sultana 2024). Quarter-quarter sections of forty acres were not only the smallest subdivision allowed for homesteading but were also the smallest used for national park units in the West. Pipe Spring NM in northwest Arizona is one such forty-acre (16 hectare) monument. Other monuments were made up of a variety of subdivisions and could take the form of an irregular shape, though still made up of straight lines and with acreage often still divisible by 40.

Like other monuments Canyon de Chelly was created by presidential proclamation, which specified the boundaries (Sullivan 1947). These were based on the Navajo Meridian and Baseline, though most of the new monuments lay outside those townships surveyed by Darling. In the absence of surveyed lands, the description given was hypothetical; only after a surveyor had marked them on the ground could they be drawn. Substantial differences could emerge between what was intended on paper and what appeared on the ground. For these reasons, it was impossible to provide a map with the proclamation, as was customary. Figure 9.4 shows a reconstruction of what these boundaries might have looked like.[2] This shows that the proclamation based on unsurveyed land missed much of the canyon system. This sort of problem was not uncommon in situations where monuments were created in advance of surveying, though this is the most extreme case known to the author.[3]

The rugged canyon country in northern Arizona does not align well with the PLSS. Navajo National Monument, on the northwest side of the reservation, is made up of two 160 acre (65 hectare) squares and a 40 acre (16 hectare) square, each centered on a cliff dwelling. The three squares are several miles apart, located in a rugged canyon and plateau

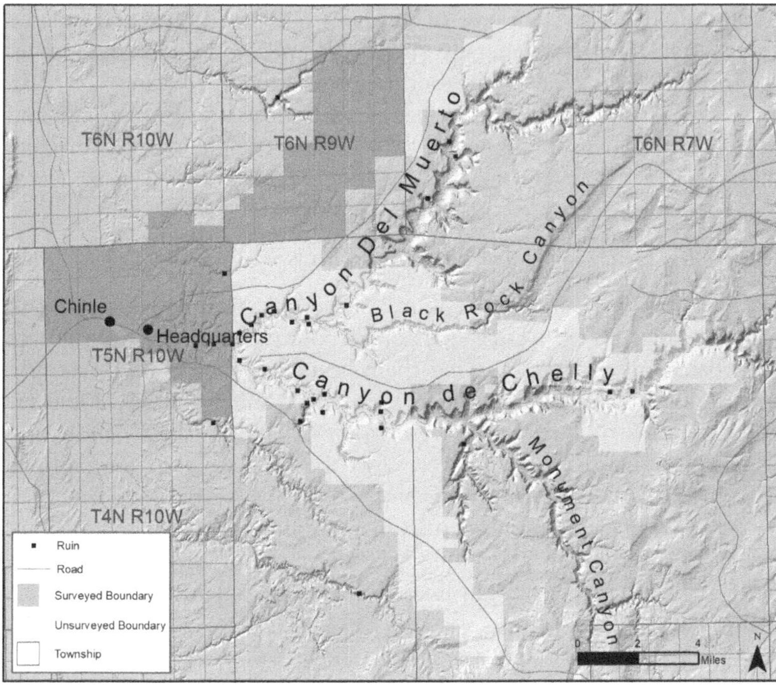

Fig. 9.4 Reconstruction of 1931 boundaries as specified in original Proclamation. *Source* Author, based on Bureau of Land Management (2020), and United States Geological Survey (2022)

topography. When the monument was created in 1909 the boundaries were considered sufficient to protect the ruins; there was no idea that the monument would later become a tourist attraction, and the boundaries were not created with this in mind (Rothman 1991). There was no room for roads within the park, and the NPS would later have to lease land from the tribe for a visitor center and campground.

Navajo NM and Mesa Verde NP show differing methods of creating a canyon park based on archaeology: either specify a small square around each ruin or create a large rectangular area to encompass everything. Canyon de Chelly's large number of ruins may have ruled out the approach used at Navajo, while the existence of a substantial resident population looked after by the Bureau of Indian Affairs may have prevented the Mesa Verde approach. Here the boundaries provided an example of a superimposed boundary, one overlaid on clearly established land uses (Dilsaver and Wyckoff 2005).

The Second Boundary

Due to several issues, including some awareness of the poor accuracy of the boundary specified in 1931 (a definite understatement), a new monument was proclaimed in 1933 (Sullivan 1947; Brugge and Wilson 1976). The west end of the canyon, in Township 5 North, Range 10 West, remained the same while the rest simply included everything within half a mile of the rim of Canyon de Chelly, del Muerto, Monument Canyon, and their tributary canyons (Fig. 9.5).[4] This solved the problem of the canyon's location not being known with certainty by largely ignoring the survey. However, this new boundary line was not surveyed or marked on the ground and would be difficult to do so, given the difficulty of defining the exact rim of the canyon. Nor was it clear why some canyons were used and not others, or why the buffered boundary took the form that it did (National Park Service, Bureau of Indian Affairs, and Navajo Nation 1989). Fortunately, there was no need to survey the boundary, as it did not mark a change in ownership or land use.

A 1937 Apache County highway map is the oldest the author has found that shows the new buffer boundary, though the west end differs from later NPS depictions (Arizona State Highway Department 1937). In contrast, the 1941 NPS tourist brochure for the National Monument appears to show the 1931 boundary, though now lined up with the canyon (National Park Service 1941). The 1947 and later brochures show the 1933 boundary (National Park Service 1947), though often at small scales. Detailed canyon maps of the period show no boundary at all. The current brochure map shows the 1933 boundary in detail, though it has never been surveyed.

National park unit boundary changes are remarkably common, with almost half of park units having had a revision and many larger units having had as many as fifteen changes (National Park Service 2016; Weber and Sultana 2024). These may be a result of the need to remedy inaccurate boundaries, expand the resources protected, or eliminate less valued lands whose management may be problematic. While the 1933 boundary revision better fit the canyons, much of the rim road mileage was still left out of the monument. In the 1950s the NPS sought an enlarged

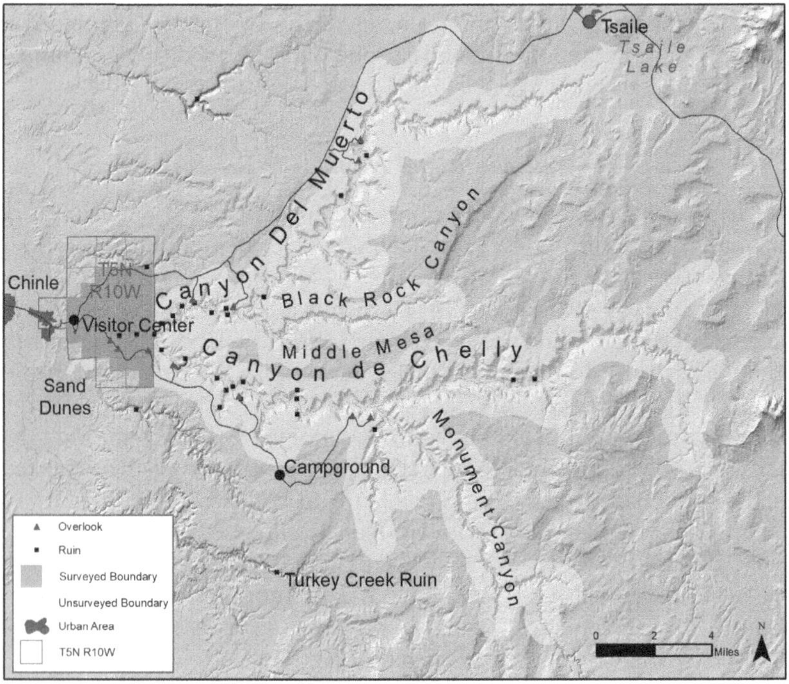

Fig. 9.5 Current boundaries of Canyon de Chelly National Monument. *Source* Author

boundary to contain the rim roads as well as avoid interstitial lands in the peninsula between the two canyons, which remained outside the monument boundary, but nothing came of this. Expanding a park's boundaries to include the entirety of the park road system has been a common strategy to ensure NPS control over the park environment and can be seen in western parks both small (Tonto NM in Arizona) and large (Dinosaur NM in Utah and Colorado) to prevent incompatible commercial development (Weber and Sultana 2024). In other cases, roads could be altered to keep them within parks; at nearby Mesa Verde National Park, the Chapin Mesa road was rerouted to avoid the Ute Mountain Indian Reservation due to a dispute about roadside development between the Tribe and the NPS (Smith 2002). While the rim roads at Canyon de Chelly frequently pass outside park boundaries, there is conversely a substantial mileage of dirt roads meandering through the monument, connecting Navajo homes or other sites within the monument. The boundaries clearly do not provide a barrier to movement.

A 1989 plan called for substantial changes to the monument's boundaries. One proposal was for boundaries based on the rim roads up to the Spider Rock and Antelope House turnoffs, and then follow the current boundaries to the east, except that Tsaile Lake (built in 1963) would be included. Another proposal was to use current boundaries but only include land below the rims in the eastern portion. Other ideas were to add the sand dunes near the headquarters and exclude Middle Mesa. These proposals did not encompass expanding the Monument to include the nearly pristine Three Turkey Ruin in the next canyon south of de Chelly, several miles outside the monument. All proposals agreed that wilderness should not be designated within the monument, as it was occupied. None of the proposals were acted upon, and the Monument boundaries remain the unsurveyed buffer specified in 1933.

The Disappearing Navajo Meridian

Beginning in 1878 the Navajo reservation was expanded to the west to include the entire Chinle valley, missed by the original proclamation. Many additional enlargements followed, and the reservation reached its present extent by 1934 (Goodman 1982). The current reservation is larger than ten states and includes most of northeastern Arizona as well as adjacent parts of New Mexico and Utah and includes much of that claimed by the tribe as part of its traditional homeland. This expansion involved treaties, Acts of Congress, presidential proclamations, and a long-running dispute over land with the Hopi tribe (whose reservation is located within that of the Navajo) not resolved until 1974 (Frantz 1999).

The Navajo survey was not expanded along with the reservation as no Navajo took up individual allotments, the original purpose of its application. The existence of the survey was also challenged by Anglo settlement along the San Juan River in northwest New Mexico. Two rows of townships were surveyed in New Mexico using that state's meridian and baseline in 1915 (Gregory 1917). These were later discovered to have had intruded on the Navajo Meridian area, though as Darling's survey markers had disappeared the exact overlap was not known (Fig. 9.6). Because no private land existed in this area it was decided to resolve the problem of overlapping townships by canceling that portion of the Navajo survey within the state of New Mexico. This took place on June 17, 1936.

Attention then turned to the Navajo survey within Arizona. It likewise had seen no use except around the headquarters area of Canyon de Chelly,

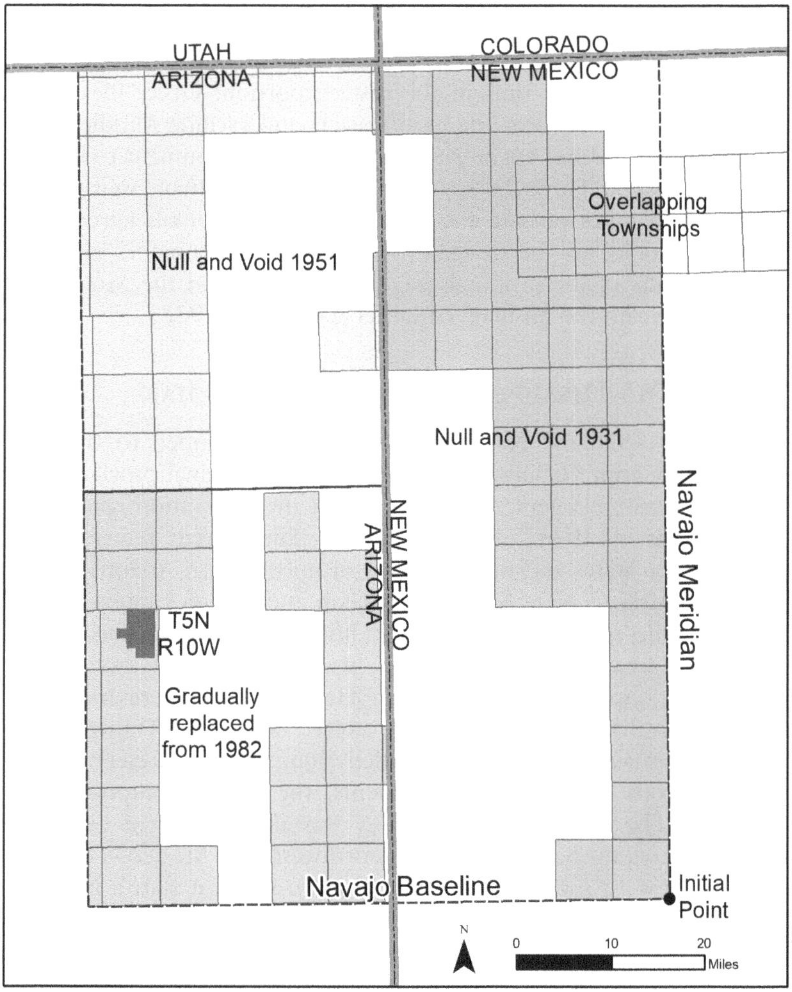

Fig. 9.6 Disappearance of the Navajo Meridian. *Source* Author

so on May 11, 1951, all townships north of Township 7 North were eliminated (White 1996). In 1982 the tribe, Bureau of Indian Affairs, and Bureau of Land Management (successor to the General Land Office) agreed to extend the Gila and Salt River meridian (which covers the rest of Arizona) over the remaining area of the Navajo survey, except for Township 5 North, Range 10 West (White 1996). A 1992 resurvey of this township was done to preserve trust land and the National Monument boundaries; this township will remain indefinitely. This current version of the township has only 15 sections, by far the smallest area (15 square miles or 3885 hectares) covered by any Public Land Survey System meridian and baseline (Fig. 9.7).[5]

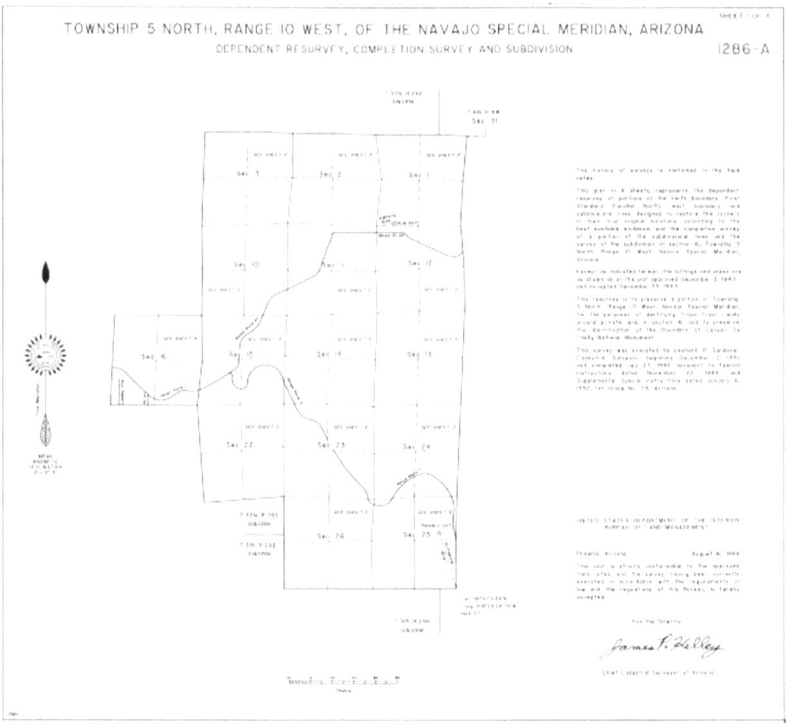

Fig. 9.7 Current survey plat for Township 5 North, Range 10 West. *Source* Bureau of Land Management (2020)

Conclusions

On a map, Canyon de Chelly NM appears to be a conventional canyon park unit, with north and south rim roads providing access to overlooks and allowing views of colored cliffs and green watercourses below. A closer inspection reveals the monument is comprised of two sections. The western or headquarters area is a traditional national park unit, with an entrance sign, visitor center, campground, and concession lodging, as can be seen in many units across the country. Most place names are in English and were bestowed by Euro-American explorers and archaeologists. This area has been surveyed and has definite boundaries. It is highly legible to the NPS as well as visitors accustomed to touring national park units and expecting to stop at the visitor center before driving a rim road to designated overlooks where they will see sights such as Spider Rock.

The remainder of Canyon de Chelly National Monument is a carefully tended cultural landscape rather than the wild canyons that might be expected in a national park unit. This is a traditional Navajo landscape, with many dispersed residences, wandering livestock, and woodcutting along the monument rim roads as well as fields and pastures on the canyon floor. Navajo place names abound but are known only by them. This part of the monument remains unsurveyed and has little contrast between those areas within and outside the monument. Although confusing to national park visitors and perhaps frustrating to NPS workers, the canyons are highly legible to the Navajo.

Most of the Monument is open only to Navajo; non-Navajo visitors are allowed only on the park's paved roads, at the ten overlooks, and the White House Ruin trail (temporarily closed since 2020), though they make take guided tours on the canyon floor. Though closed to visitors it is not a traditional backcountry due to the presence of residents, homes, and fields. For that reason, the divisions between the west end and canyon sections do not correspond to traditional front and backcountry areas in a national park. Nor is there any wilderness. Monument visitors will catch glimpses of traditional native life; Navajo residents will frequently see tourists, but the two groups are unlikely to interact.[6] For all intents and purposes they are not part of the same area.

The monument's rationale was archaeology, but the ruins had to be viewed from overlooks on the canyon rim due to the need to keep roads out of canyons. With undeveloped canyons and spectacular views, the monument has become a scenery park similar to others on the Colorado

Plateau. Newer monuments in the region, such as Bears Ears or Grand-Staircase Escalante do not have a similar level of development, unity of features, or legibility. Bears Ears is also contested by Euro-American settlers who see Navajo claims on this area as inappropriate since it lies outside the Navajo Reservation and therefore not part of Navajo territory (Roberts 2021).

Underlying these boundaries remains the fact that the land within the monument, however bounded, remains the property of the Navajo Nation. This is not the case at Navajo NM and Hubbell Trading Post National Historic Site, which are on land owned by the NPS. The monument remains one of two with a resident Native American population (the other being Death Valley National Park, home to the Timbisha Shoshone Indian Reservation). Canyon de Chelly is an exception to the NPS' generally dismal record with Native Americans (Keller and Turek 1998; Spence 1999; Burnham 2000). While inholdings, or non-federal land, are common within national park units, the only comparable park unit is the National Park of American Samoa, which is leased from villages. In this territory, never surveyed for the PLSS, the land was traditionally in communal ownership.

Although Canyon de Chelly does not rank as a well-known park unit, its existence is significant because it represents the park that most strongly departs from the 'Yellowstone model' or 'Fortress Yellowstone' approach to park development followed by the NPS from the inception of the park system (Weber and Sultana 2024). This model has been rejected as unworkable or morally dubious by most other countries, many of which have national parks that contain resident populations and their fields and may also be co-managed with these people. This alternate model is sometimes known as the 'Uluru model' after the park in Australia (Adams 2014). It was first sighted by Euro-Australians in 1873 and became a popular tourist attraction, though remaining a sacred place to Aboriginal residents. They now have an active voice in managing the park.

Just as the arrival of U.S. government forces in 1864 shattered the Navajo's idea of the canyons as a fortress, the persistence of the Navajo may help shatter the National Park Service's view of parks as fortresses. Canyon de Chelly is the closest park to the 'Uluru model' in the U.S. park system. However, it appears that park management has so far followed traditional NPS policies. For example, a 2010 vegetation mapping project (Thomas et al. 2010) was carried out mainly by NPS staff and relied on biological conventions for naming vegetation. Those Navajo living in the canyon were thanked for their patience during fieldwork but were not consulted. The NPS may also not close places sacred to Native Americans because of constitutional restrictions on favoring any religion. This has been a major point of contention between tribes and the NPS at park units such as Rainbow Bridge and Devils Tower, but the fact that most of Canyon de Chelly is closed to non-Navajo may have helped reduce such tensions here. There are clearly many possibilities for more cooperative management.[7]

Finally, it is interesting that the reservation, the survey, and the monument have had many influences on one another. The Navajo Reservation was originally defined by the location of Canyon de Chelly. Had there been no Navajo Reservation there would have been no need for a Navajo Meridian and Baseline, and the monument would likely have been created decades earlier. Had there been no Navajo Meridian and Baseline the boundaries of Canyon de Chelly NM would now be different. Had there been no Canyon de Chelly National Monument the Navajo Meridian and Baseline would no longer exist.[8] Though the interests and motives of the Diné, land surveyors, and National Park Service were quite different, the canyon was central to the places created by each. Even more remarkably, Canyon de Chelly marks the point of beginning for the modern topographic mapping of the United States: in 1886 Canyon de Chelly was included on the first topographic map sheet published by the United States Geological Survey or USGS (Evans and Frye 2009; Weber 2024).

Declarations

Conflict of interest The author has no conflict of interest to declare that are relevant to the content of this chapter.

Notes

1. de Chelly is a Spanish approximation of the Navajo word Tséyi', which simply means 'canyon.' It is pronounced 'duh-Shay.' Canyon de Chelly is used to refer both to one of the two main canyons as well as the entire canyon complex.
2. Maps of the Navajo Meridian are rare. Aside from GLO plats the Arizona portion of the Navajo meridian appears on a 1937 Apache County highway map (Arizona State Highway Department 1937) and the 1955 Canyon Del Muerto, Yellowstone Canyon, Nazlini,

Lukachukai, Sonsela Buttes, and Buell Park 15' topographic maps produced by the United States Geologic Survey (2022). The only map the author has found (other than GLO plats) showing the New Mexico portion of the survey is in Gregory (1917).

3. Great Sand Dunes NM in Colorado was proclaimed in an area not yet fully surveyed. Its initial boundaries in 1932 were later found to include all or parts of eight sections that did not in fact exist (Geary 2016).
4. The NPS had sought a further expansion to the west but this did not happen. The General Land Office conducted a new survey of Section 15 of Township 5 North, Range 10 West in 1933 and 1943 to help clarify the location of thew western end of the monument boundary.
5. The standard map of principal meridians and baselines put out by the Bureau of Land Management shows the Arizona portion south of Township 7 North, vastly exaggerating its area. The author has seen no maps of the PLSS that shows the correct size of the Navajo survey after it was reduced.
6. The most recent visitor survey (National Park Service 2012) shows only three percent of monument visitors are Native American, and there is no way to identify how many are Navajo. A survey at Rainbow Bridge National Monument on the western edge of the reservation showed that two percent of visitors were Native American, with Navajo as one of six tribal affiliations given (National Park Service 2007).
7. The first Native American director of the NPS was appointed in 2021. In an interview, he stated that tribal co-management of national park units would not change the visitor experience, though it should help long-term management of plants and animals (Schechter 2022).
8. Not only was Darling's survey of the Navajo meridian almost entirely erased from the map, but most physical traces of it disappeared as well. The Initial Point for the Navajo Meridian and Baseline was lost for many years. White (1996) helped locate it, at 35° 45′ 06.71″ North latitude, 108° 32′ 12.15″ West longitude. This and the west boundary of Canyon de Chelly NM are the sole surviving remnants of Darling's survey work in the west. His other survey work in the west fared even worse. His 1870 survey of the Chickasaw lands in Indian Territory (now Oklahoma), for which he established the

Indian Meridian and Baseline, was declared null and void in 1897. His 1868 survey of the New Mexico-Colorado state line was challenged in court due to errors, and although upheld in 1925 was later resurveyed (Cazler 1976).

References

Adams, M. 2014. Pukulpa pitjama Ananguku ngurakulu—Welcome to Anangu Land: World Heritage at Uluru-Kata Tjuta National Park. In *World heritage sites and indigenous people's rights*, ed. Stefan Disko and Helen Tugendhat, 289–311. Copenhagen: IWGIA.

Andrews, T.J. 1991. Ecological and historical perspectives on Navajo land use and settlement patterns in Canyons de Chelly and Del Muerto. *Journal of Anthropological Research* 47: 39–67.

Arizona State Highway Department. 1937. *General Highway and Transportation Map, Apache County, Arizona*. Phoenix: Arizona State Highway Department.

Brugge, D.M., and R. Wilson. 1976. *Administrative history: Canyon de Chelly national monument, arizona*. Washington, DC: National Park Service.

Bureau of Land Management. 2020. *General land office records*. Available at https://glorecords.blm.gov/default.aspx

Burnham, P. 2000. *Indian country, god's country: Native Americans and the national parks*. Washington, DC: Island Press.

Cazler, L. 1976. *Surveys and surveyors of the public domain, 1785–1975*. Washington, DC: Government Printing Office.

Dilsaver, L.M., and W. Wyckoff. 2005. The political geography of national parks. *Pacific Historical Review* 74: 237–266.

Evans, R. T., and H. M. Frye. 2009. *History of the Topographic Branch (Division)*. USGS Circular 1341.

Feldman, J.W. 2011. The need for legible landscapes: Environmental history and NPS management at Apostle Islands National Lakeshore. *George Wright Forum* 28 (2): 148–160.

Frantz, K. 1999. *Indian reservations in the United States: Territory, sovereignty, and socioeconomic change*. Chicago: University of Chicago Press.

Geary, M.M. 2016. *Sea of sand: A history of great sand dunes national park and preserve*. Norman: University of Oklahoma Press.

Goodman, J.M. 1982. *The Navajo atlas*. Norman: University of Oklahoma Press.

Grant, C. 1978. *Canyon de Chelly: Its people and rock art*. Tucson: University of Arizona Press.

Gregory, H.E. 1917. *Geology of the Navajo Country: A Reconnaissance of Parts of Arizona, New Mexico, and Utah*. USGS Professional Paper 93.

Harmon, D., F.P. McManamon, and D.T. Pitcaithley, eds. 2006. *The antiquities act: A century of american archaeology, historic preservation, and nature conservation*. Tucson: University of Arizona Press.
Hart, E.R. 2008. The west boundary of the Hualapai Indian Reservation. *Western Legal History* 21: 121–163.
Hubbard, B., Jr. 2009. *American boundaries: The nation, the states, the rectangular survey*. Chicago: University of Chicago Press.
Iverson, P. 2002. *Diné: A history of the Navajos*. Albuquerque: University of New Mexico Press.
Jenkins, J. 2018. A 'deep'aesthetics of contested landscapes: Visions of land use as competing temporalities. *Geoforum* 95: 35–45.
Jett, S.C. 1978. The origins of Navajo settlement patterns. *Annals of the Association of American Geographers* 68: 351–362.
Jett, S.C. 1993. An introduction to Navajo sacred places. *Journal of Cultural Geography* 13: 29–39.
Jett, S.C. 1994. Physical characteristics of Navajo trails, Canyon de Chelly area, Arizona. *Material Culture* 26: 37–48.
Jett, S.C. 2001. *Navajo placenames and trails of the Canyon de Chelly System, Arizona*. New York: Peter Lang.
Kappler, C.J. 1904. *Indian affairs laws and treaties*, vol. II. Washington, DC: Government Printing Office.
Keller, R.H., and M.F. Turek. 1998. *American Indians and national parks*. Tucson: University of Arizona.
Kessell, J.L. 1981. General Sherman and the Navajo Treaty of 1868: A basic and expedient misunderstanding. *Western Historical Quarterly* 12: 251–272.
Linford, L.D. 2000. *Navajo places: History, legend, landscape*. Salt Lake City: University of Utah Press.
Lynch, K. 1960. *The image of the city*. Cambridge: MIT Press.
Mackintosh, B., J.A. McDonnell, and J.H. Sprinkle Jr. 2018. The national parks: Shaping the system, fourth edition. *George Wright Forum* 35: 2–132.
McPherson, R.S. 1992. *Sacred land, sacred view: Navajo perceptions of the four corners region*. London: Charles Redd Center for Western Studies, Brigham Young University.
National Park Service. 1941. *Canyon de Chelly national monument, Arizona*. Washington, DC: Government Printing Office.
National Park Service. 1947. *Canyon de Chelly national monument*. Washington, DC: Government Printing Office.
National Park Service. 2007. *Rainbow bridge national monument visitor study*. Park Studies Unit Visitor Services Project Report 194.
National Park Service. 2012. *Canyon de Chelly national monument visitor study*. Natural Resource Report NPS/NRSS/EQD/NRR—2013/730.

National Park Service. 2016. *The national parks: Index 2012–2016*. Washington, DC: National Park Service.
National Park Service, Bureau of Indian Affairs, and Navajo Nation. 1989. *Joint Management Plan for Canyon de Chelly National Monument*. Washington, DC: National Park Service.
Pearce, M.W. 2004. The holes in the grid: Reservation surveys in Lower Michigan. *Michigan Historical Review* 30: 135–166.
Roberts, D. 2021. *The Bears Ears: A Human History of America's Most Endangered Wilderness*. New York: W. W. Norton.
Roeder, F. 2009. Ehud N. Darling. *LIDAR magazine*. https://lidarmag.com/2009/03/21/ehud-n-darling/
Rohn, A.R., and W.M. Ferguson. 2006. *Puebloan ruins of the southwest*. Albuquerque: University of New Mexico Press.
Rothman, Hal. 1991. *Navajo national monument: A place and its people*. Washington, DC: National Park Service.
Rothman, H. 1994a. *America's national monuments: The politics of preservation*. Lawrence: University Press of Kansas.
Rothman, H. 1994b. *Preserving different pasts: The American national monuments*. Urbana: University of Illinois Press.
Runte, A. 2010. *National parks: The American experience*, 4th ed. Lanham: Taylor Trade Publishing.
Sack, R.D. 1986. *Human territoriality: Its theory and history*. Cambridge: Cambridge University Press.
Scott, J.C. 1998. *Seeing like a state: How certain schemes to improve the human condition have failed*. New Haven: Yale University Press.
Sellars, R.W. 2007. A very large array: Early federal historic preservation—The Antiquities Act, Mesa Verde, and the National Park Service Act. *Natural Resources Journal* 47: 267–328.
Schechter, A. 2022. Meet the new man behind the National Park Service. *New York Times*. https://www.nytimes.com/2022/06/24/travel/national-park-service-director.html. Last accessed July 18, 2022.
Smith, D.A. 2002. *Mesa Verde national park: Shadows of the centuries*. Boulder: University Press of Colorado.
Spence, M.D. 1999. *Dispossessing the wilderness: Indian removal and the making of the national parks*. Oxford: Oxford University Press.
Sullivan, T.A. 1947. *Proclamations and orders relating to the national park service up to January 1, 1945*. Washington, DC: Government Printing Office.
Thomas, K.A., M.L. McTeague, L. Ogden, K. Schulz, T. Fancher, and A. Cully. 2010. *Vegetation classification and distribution mapping report*. Natural Resource Technical Report NPS/SCPN/NRTR-2010/306.
United States Geological Survey. 2022. *TopoView*. https://ngmdb.usgs.gov/topoview/

Weber, J. 2021. The lost meridian of the 'Aha Makhav. *Cartographica* 56: 106–119.

Weber, J. 2023. The public land survey landscape in the intermountain American west. *Professional Geographer* 75 (3): 452–462.

Weber, J. 2024. The unfinished wheeler atlas of the American West. *Imago Mundi.*

Weber, J., and S. Sultana. 2024. *The parks belong to the people: The geography of the national park system.* Athens: University of Georgia Press.

Webster, G.R., and J.I. Leib. 2011. Living on the grid: the U.S. rectangular public land survey system and the engineering of the American landscape. In *Earth engineering*, ed. Stan Brunn, 2123–2138. Dordrecht: Springer.

White, C.A. 1983. *A history of the rectangular survey system.* Washington, DC: Government Printing Office.

White, C.A. 1996. *Initial points of the rectangular survey system.* Westminster, CO: Professional Land Surveyors of Colorado Inc.

Winslow, H. 2013. 'Strangers on their own land': Ideology, policy, and rational landscapes in the United States, 1825–1934. *Cartographica* 48: 47–66.

CHAPTER 10

Youth Visitor Trends in United States National Parks: A Preliminary Study

Matthew Balentine, John Crowell, Charles F. Holloman III, and Michael Pretes

Introduction

"In 2014, America's national parks attracted a record-setting 292.8 million visits, but the typical visitor to the country's biggest parks is edging closer to retirement age." So observed the media site CNN in 2015, suggesting that the number of younger national park visitors in the

M. Balentine · J. Crowell · C. F. Holloman III · M. Pretes (✉)
Department of Geoscience, One Harrison Plaza, University of North Alabama, Florence, AL 35632, USA
e-mail: mjpretes@una.edu

M. Balentine
e-mail: mdbalentine@una.edu

J. Crowell
e-mail: jcrowell@una.edu

C. F. Holloman III
e-mail: cholloman@una.edu

J. Weber and S. Sultana (eds.), *The Changing Geography of National Parks and Protected Areas*,
https://doi.org/10.1007/978-3-031-74653-6_10

United States is declining. Likewise, the British newspaper *The Economist* noted in 2013 that “America’s national parks struggle to attract young visitors.” Though these quotes are 10 years old, recent academic studies such as Ruiz and Havlick (2023) also express concern about younger generations and their interest in visiting national parks. Is the United States National Park System doomed to insignificance as the population ages? Are younger Americans more interested in online recreational experiences and not in visiting national parks? Our study used actual observation and counting of park visitors, supplemented with ranger and park staff interviews and a survey, to assess national park population dynamics in the United States. Our results show that younger visitors are visiting national parks in significant numbers.

As noted above, recent media reports have suggested that members of the “Millennial Generation” (those born between 1981 and 1996) and “Generation Z” (those born between 1997 and 2012) are less inclined to visit national parks, compared to older generations. Our research, conducted over six years at eight national parks in the United States, aimed to determine whether the number of Millennial and Generation Z visitors to national parks is truly declining, and, if so, why. Are young people interested in visiting national parks, or are competing technology-based recreational attractions, such as video games and other computer-based activities, leading to a decline in younger park visitors? Data on the age structure of national park visitors are not readily available (the National Park Service has only recently begun to collect this information in any meaningful way, and only for a few selected locations), and media reports are largely based on anecdotal evidence. In our study, we collected quantitative and qualitative data at eight national parks—Olympic, Mount Rainier, Crater Lake, Redwood, Great Basin, Grand Canyon, Petrified Forest, and Big Bend—in the western states of Washington, Oregon, California, Nevada, Arizona, and Texas. These eight parks served as case studies to determine the trends in Millennial and Generation Z visitation and the factors accounting for these trends, and to assess generational recreational preferences.

The United States population in 2014 is estimated at 342 million and is projected to grow to about 383 million in 2054 (largely due to immigration) and then decline by the end of the century (United States Census Bureau 2023, United States Congressional Budget Office 2024). According to the same sources, the population is expected to age, with those over 65 years old outnumbering those under 18 by the mid to late century. It is common in the United States to classify the population by generations, each identifiable by a range of birth years and then by alleged common cultural features and attitudes. Though generational classifications and the assumptions that go with them can be vague and problematic, such categories have entered common media discourse and are widely used (see, for example, the study by Twenge 2023). Overall, and in a general sense, generational classification can give an overview of attitudes toward society, culture, politics, and style (among other things), including attitudes toward recreation. According to the statistics site Statistica.com (2024), the United States population consists of seven generations listed from oldest to youngest: Greatest Generation (born before 1928), Silent Generation (born 1928–1945, Baby Boomer Generation (born 1946–1964), Generation X (born 1965–1980), Millennial Generation (born 1981–1996), Generation Z (born 1997–2012), and Generation Alpha (born after 2012). Table 10.1 indicates the percent of each generation in the United States population:

In our study, we classify the Millennial Generation, Generation Z, and Generation Alpha populations as "younger" generations and the others as "older" populations and identified and aggregated them this way in our visitor counts at national parks.

A recent debate in recreation studies and environmental geography is whether technology-driven recreation, such as video games, social media, and activities like Pokémon Go, is reducing the number of younger visitors to national parks in the United States. National Park Service data indicate that visitation to national parks is increasing, but these data do not usually reveal the actual age of the visitors. The use of electronic devices for recreation has been identified as a primary factor in the decline in visitation of United States national parks since 1988 (Pergams and

Table 10.1 Percent of U.S. population by generation, 2024

Generation	*Birth Years*	*Percent (%)*
Greatest generation	Before 1928	0.13
Silent generation	1928–1945	4.92
Baby boom generation	1946–1964	20.93
Generation X	1965–1980	19.51
Millennial generation	1981–1996	21.71
Generation Z	1997–2012	20.69
Generation Alpha	After 2012	12.76

Zaradic 2006, 2008). The studies of Pergams and Zaradic document a decline in total park visitors per capita but do not include data on the ages of visitors. Though their studies suggest that technology-driven recreation is a leading factor in visitation decline, they say nothing about whether this is especially true for Millennials and Generation Z. A report in *The Economist* (2013) notes that although data are scant, in Great Smoky Mountains National Park the share of summer visitors aged 61 or over rose 10% during a survey conducted in 1996 and increased by 17% in a similar survey in 2008; the share of those 15 and under fell from 26 to 22%. Little research has been done on the age of visitors since 2008, and there is no clear quantitative data to support a decline in Millennial and Generation Z visitors. And yet this issue is important, for as Hungerford and Volk (1990) and Duda et al. (1998) have noted, exposure to the outdoors in such places as national parks is crucial in educating young people about the environment and the human impacts on it.

The most recent literature on this subject includes the work of Xiao et al. (2022) and Ruiz and Havlick (2023). The former study is based on a very large survey of over four thousand respondents, but the focus is on the racial and ethnic backgrounds of visitors to determine whether diverse audiences are visiting national parks (which is not a factor in our project). Very little is said about visitor ages in this study. In the work of Ruiz and Havlick (2023) the focus is on the Millennial Generation and its knowledge, awareness, and attitudes toward national parks. The authors found that Millennials often lack knowledge of the national park system, but still indicate strong support for it; little is said about overall age dynamics including Generation Z as well as Millennials. A very recent book by Weber and Sultana (2024), while not specifically addressing the question of visitor age dynamics, does note that some of the more remote parks, such as those in Alaska, tend to see more older visitors because of the increased time and cost of visiting these parks, given that they are remote from population centers and in most cases require air travel to reach Alaska.

The National Park Service is beginning to recognize the need for demographic data including visitor ages and to that end, its visitor surveys have begun to incorporate more assessment of visitor age dynamics. Yet this coverage is done largely by voluntary questionnaires which do not assess a randomly and non-self-selected visitor population, and many of the surveys are very old (see Washington State University 1982–2017). Moreover, not all parks are covered. Of our eight parks, only five have conducted surveys that ask about visitor ages: Olympic in 2000, Mount Rainier in 2012, Crater Lake in 2001, Redwood in 1993, and Big Bend in 1992. The most recent of these reports is 12 years old and some date from more than 30 years ago.

More recently, a large-scale general assessment of park visitors was completed by the National Park Service (2023). This report covers many aspects of the visitor experience, including a question on "party characteristics" or composition of the visiting group. The report breaks down visitors into adults and children (under 18 years old) and does ask about the ages of visitors (individual reports available for some parks also provide this data in disaggregated form for each park covered). The parameters used were nine categories that do not align with generations, so generational analysis is difficult. Furthermore, this data was collected by voluntary surveys containing 39 questions, so we can assume that many people refused to answer, did not complete or return the survey, did not understand the questions because of language, or gave inaccurate or false information (a common occurrence when asking about people's ages). The intercept response rate was 84% overall but the post-trip survey response rate was only 33%. Though it is encouraging to see the National Park Service begin to collect data about the ages of park visitors, the methods used may not take into account the larger population of park visitors compared to a direct counting of all visitors at selected times and places, especially as they focused only on the busiest times at the parks and none of the eight parks used in our study was included in this National Park Service assessment.

Selected National Parks

Of the 63 national parks in the United States (at present count), we selected eight of them, based on our own experience, knowledge, and connections to these places. We also ensured that the selection of eight cases represented large and small parks, those with many and those with few annual visitors, those located near large urban areas and those more remotely located, and with a variety of natural landscapes and attractions. This careful choice of parks avoids selection bias and allows for a diversity of parks to be represented in our analysis. Though the parks selected are all in the western part of the country (the Pacific Northwest and the Southwest in particular), they are located within six different states, and they nevertheless are a diverse enough selection for our project. Some of these parks have high annual visitations (Olympic, Grand Canyon) while some have relatively few visitors (Redwood, Great Basin, Big Bend). These eight parks also reflect a diversity of landscapes, ecosystems, and historical and cultural resources including mountain, rainforest, coastal, and desert environments, Native American and settler cultural artifacts and resources, and varying recreational opportunities (Fig. 10.1).

Looking at these parks from a general northwest to southeast direction, we begin with *Olympic National Park* located in Washington State in the far northwest of the conterminous United States. Olympic is one of the older U.S. national parks, having been established in 1909, and is 1442 square miles [3735 square kilometers] in size. Its natural environment is extremely diverse, incorporating high mountains reaching 7980 feet [2428 m], extensive temperate rainforest, and part of the Pacific Ocean coastline. Olympic has several well-developed visitor facilities and

Fig. 10.1 Map showing location of the eight national parks used in this study (*Author* John Crowell)

is approximately a two-hour drive from the Seattle/Tacoma metropolitan area with its over four million people, and consequently has a high annual visitation of over 2.9 million people (in 2023) (irma.nps.gov/stats/reports/park/OLYM). We conducted visitor counts at all three park visitor centers: Olympic National Park Visitor Center (in Port Angeles), Hurricane Ridge Visitor Center, and Hoh Rain Forest Visitor Center, as well as at Kalaloch Ranger Station (which contains a small visitor area). The methods used for counting are explained in the Methods section, below.

The second park in this study is *Mount Rainier National Park*, also located in Washington State, which is just under a two-hour drive from the Seattle/Tacoma metropolitan area. Mount Rainier received almost 1.7 million visitors in 2023 (irma.nps.gov/stats/reports/park/MORA). Mount Rainier is also one of the older parks, having been established in 1899, and centers around the massive dormant stratovolcano of Mount Rainier, which reaches an elevation of 14,410 feet [4390 m] and contains 25 glaciers. The park is 369 square miles [957 square kilometers] in area. We conducted visitor counts at all three visitor centers, including the main one, Henry M. Jackson Memorial Visitor Center located at Paradise, as well as at Sunrise Visitor Center and Ohanapecosh Visitor Center. We also conducted counts at the Longmire Museum (which also functions as a visitor center) as well as on several trails in various regions of the park.

Crater Lake National Park in Oregon is the third national park in our study. It is also one of the older parks, having been established in 1902. The park has a remote location, with no large city located nearby, and is small, only 286 square miles [741 square kilometers]. The nearest regional airport is at Medford, about a two-hour drive away, and the state's largest city, Portland, is almost four hours away by car. The park centers around the eponymous Crater Lake, the caldera remnant of the ancient Mount Mazama. It is the deepest lake in the United States at 1949 feet [594 m] deep and its waters are exceptionally clear, giving the lake a bright blue color. As a caldera, the lake's edges are exceptionally steep, with only one point of access to the shoreline by a steep but paved road. This road accesses the ramp for boats to Wizard Island, a small conical volcanic island within the lake. Given its remote location but distinctive and photogenic beauty, the park received approximately 500,000 visitors in 2023 (irma.nps.gov/stats/reports/park/CRLA). The park contains two visitor centers, the Rim Visitor Center and the Steel Information Center, at both of which we conducted visitor counts. We also conducted counts on various trails including the one to the summit of Wizard Island.

Our fourth national park is *Redwood National Park* located in the far northwest corner of California. Redwood is one of the more remote national parks in the conterminous United States and is not easily accessible from any large or mid-size city. It can only be accessed by small and winding highways or through the regional airport at Eureka (which has unpredictable weather and is frequently closed due to fog). The park is commonly known as Redwood National and State Parks as this larger unit contains the national park as well as three California state parks and effectively functions as one park unit under dual administration, covering about 217 square miles [563 square kilometers]. The park was established in 1968 to preserve old-growth groves and stands of the coast redwood (*Sequoia sempervirens*), the world's tallest species of tree, which were threatened by logging. Today the park complex protects almost half of the old-growth redwoods in the world—some of which are more than 2000 years old—in a unique forest environment; the park also includes a coastal area as well. Redwood National Park received just over 400,000 visitors in 2023 (irma.nps.gov/stats/reports/park/REDW). We counted visitors at the Park Information Center/Park Headquarters in Crescent City, Prairie Creek Visitor Center, Kuchel Visitor Center, and Hiouchi Visitor Center, as well as on various trails.

Leaving the Pacific Northwest for the desert Southwest of the United States, our fifth national park in this study is *Great Basin National Park* in Nevada. Great Basin is both a newer park (established in 1986) and one of the most remote in the conterminous United States, being more than a four-hour drive across a desert (and deserted!) road from the nearest major airport at Salt Lake City in Utah. It is located near the town of Baker (population 60 people) just off U.S. Highway 50, often known as the "Loneliest Road in America." The park protects ancient Bristlecone Pines (*Pinus longaeva*), some of which are over 5000 years old; part of the Great Basin Desert ecosystem; Mount Wheeler (13,063 feet [3982 m]; and Lehman Caves. Great Basin National Park is a designated Dark Sky Park and uses the motto "Half the Park is After Dark." Given its remoteness, the park received just over 140,000 visitors in 2023 (irma.nps.gov/stats/reports/park/GRBA). The park is quite small at 121 square miles [313 square kilometers]. We conducted visitor counts at both the Great Basin Visitor Center and Lehman Caves Visitor Center, as well as on various trails.

Our sixth national park is one of the best known in the world: *Grand Canyon National Park* in Arizona. This park, established in 1919, protects much of the Grand Canyon of the Colorado River, a spectacular gorge known for its massive size, depth, and the distinctive coloration of its geological strata. The park is a major tourist destination within the United States and attracted over 4.7 million visitors from all over the world in 2023 (irma.nps.gov/stats/reports/park/GRCA). The park is 1904 square miles [4931 square kilometers] in area. The park is approximately a one-and-a-half-hour drive from the regional center of Flagstaff, Arizona and under three hours from Las Vegas, Nevada (population 2.2 million). Most park visitors visit the area south of the Colorado River known as the South Rim and this is where we conducted our visitor counts, specifically at the Grand Canyon Visitor Center (the primary visitor center in the park) and the Desert View Visitor Center as well as on trails.

The nearest other national park to the Grand Canyon is *Petrified Forest National Park* in northeast Arizona, which is our seventh park. This park is known for its extensive deposits of fossilized petrified wood, sometimes considered a forest (though all the petrified trees are lying down, sometimes to the surprise of visitors). The park is 346 square miles [896 square kilometers] in area. The park is also in the region known as the "Painted Desert" for its colorful stones and geologic formations. The park was established in 1962 and it is located along U.S. Interstate Highway 40, a major east–west thoroughfare in the southern part of the country. Many park visitors are thus stopping on the way to somewhere else (such as the Grand Canyon) and the park is not generally seen as a distinctive destination in its own right. Nevertheless, the park received almost 800,000 visitors in 2023 (irma.nps.gov/stats/reports/park/PEFO). We conducted visitor counts at the Painted Desert Visitor Center, Painted Desert Inn (an historic hotel), and the Rainbow Forest Museum, as well as on trails.

Our eighth and final national park is *Big Bend National Park* in Texas. The park is located in a large bend in the Rio Grande/Rio Bravo River in west Texas and it borders Mexico. It is 1252 square miles [3243 square kilometers]. The park contains exceptional desert scenery and part of the Chihuahuan Desert ecosystem, with its distinctive flora and fauna. Like Great Basin, Big Bend is also designated as a Dark Sky Park. Big Bend was established as a national park in 1944. It is extremely remote from population centers, being more than four hours away by car from El Paso,

Texas, and Midland/Odessa, Texas, the two nearest major cities. The park received just over 500,000 visitors in 2023 (irma.nps.gov/stats/reports/park/BIBE). We conducted visitor counts at Panther Junction Visitor Center, Chisos Basin Visitor Center, Rio Grande Village Visitor Center, and Castolon Visitor Center, as well as at Chisos Mountain Lodge (the only lodge within the park) and on various trails.

Collectively these eight national parks represent various facets of the 63 United States national parks. Some (Olympic, Mount Rainier) are near major population centers, while some (Redwood, Great Basin, Big Bend) are extremely remote. Some are large in size (Grand Canyon, Big Bend) while some are much smaller (Crater Lake, Redwood, Great Basin). Some are internationally renowned (Grand Canyon, Redwood, Mount Rainier) while some are less well known (Great Basin, Petrified Forest, Big Bend).

Methods

We used three methods to collect data about Millennial and Generation Z visitation to national parks. First, we conducted a 12-question survey using Qualtrics survey tools and distributed through social media including Facebook and X/Twitter, to determine what kinds of recreational activities are most popular, and to assess the level of interest in survey respondents in visiting national parks. We collected 140 responses, the majority of which were from the Southern states of the United States, but with many from other regions as well as from other countries. The survey was approved by our university Institutional Review Board. We then conducted fieldwork at the eight national parks mentioned above. These were selected because, though all are in the western United States, they have different geographical features, different amenities, different visitor numbers, and some parks are close to a large city, and some are not. Thus, these eight cases represent most of the differing components of national parks and therefore avoid any selection bias in the choice of cases. During these visits, we interviewed park rangers and operators of local businesses to assess their perceptions of visitor age trends (the survey questionnaire was also approved by our university Institutional Review Board). These semi-structured interviews were our second research method and resulted in substantial qualitative longitudinal data on visitors. A total of 38 people were interviewed, including 24 park rangers, 2 park superintendents, and 12 local business owners.

We also counted visitors (our third research method) at selected visitor centers and at other locations, including trails and parking lots, in these eight parks (at different times of day and scattered over several days in different seasons in the years 2017–2023) and identified the number of visitors assessed by estimates of their ages as in the Millennial or Generation Z categories. This resulted in extensive quantitative data about the numbers of park visitors and their ages that allows for comparison across the different parks as well as providing overall data about visitor age trends. In all cases, more than one of the authors was present, so count totals were averaged to avoid any age bias (the ages of authors during counting periods ranged from 21 to 60). Visitors were observed from a distance in a non-interactive, non-intrusive way and counted, with the authors categorizing each visitor by either a Generation Z, Millennial, or older generation estimate.

Findings

Results suggest that younger visitors (Millennials and Generation Z) continue to visit national parks in large numbers. Aggregated data from visitor counts across the eight parks and over the six years of study found that 32% of park visitors were Generation Z, 30% Millennial, and 38% "older" generations (with Generation Alpha being lumped into Generation Z for the purposes of this study). Younger visitors (Millennials and Generation Z) constituted a total of 62% of park visitors, compared to constituting 42% of the United States population overall (55% if Generation Alpha is included). These results, based on counting of visitors and estimating their ages at visitor centers and trails across eight national parks, suggest that younger visitors are continuing to visit national parks in substantial numbers.

In our online survey, when asked what their preferred recreation option was, 12% of younger visitors identified visiting a national park as their first choice (behind beach, big city, and cruises, but above resorts and "other"). When asked why they didn't visit a national park, only 10% said they were not interested. Reasons given for inability to visit included not enough time, too far away, lack of information, and too expensive (see Table 10.2). When asked whether they would prefer to visit a national park, play video games, watch Netflix/Hulu, or engage in extensive social media, 70% of younger visitors said they would prefer to visit a national park.

Table 10.2 Responses to the question "Why don't you visit a national park?"

Response	*Percent (%)*
Not enough time	37
Too far away	23
Lack of information	13
Too expensive	12
Too crowded	5
No interest	10

Interviews with park rangers and local business owners supported these results and confirmed the continued interest of younger people in visiting national parks. Most of these respondents stated that in their experience at that park or in the surrounding area, they saw "no decline" in the number of younger visitors over time. Many respondents also stated that they witnessed a "perfect balance" of younger and older people. These respondents also noted the seasonality of visitors, with more younger visitors in summer. These, more anecdotal, responses confirm visitor count and survey data and suggest that a decline in younger visitors to national parks is not a perception or concern of park staff as well as of local business owners.

Conclusion

Overall, our results suggest that younger visitors continue to visit national parks in large numbers. Moreover, video games and online activities *do not* seem to be a major deterrent to national park visitation; the largest deterrents for younger visitors were lack of time and lack of proximity. Disaggregated data further suggest that proximity to a large city was not a factor in the number of young visitors to national parks. Our research into youth visitor dynamics continues and will incorporate more parks spread over a wider geographic area, repeat counts, and increased variation in seasonality. Finally, some interview results indicate that park visitors may be bypassing visitor centers and instead are using only online resources for park information, meaning that they may not have appeared in these counts (and are more likely to be younger). This last point is to be taken up in continuing research. The upshot of our work so far is that younger people still visit national parks in the United States in substantial numbers and are likely to continue to do so in the future. This gives hope for the increased protection of our country's public spaces and for an increase in the number of units in the system representing a larger diversity of America's natural and cultural wonders.

Acknowledgements We thank all the park superintendents, rangers, staff, volunteers, and local business owners and employees who were interviewed for this study. We thank Ethan Lacher and Hayden Wilkes for helping to count visitors at Petrified Forest National Park and Chase Mathis for helping to count visitors at Grand Canyon National Park. We thank the University of North Alabama Department of Geoscience Freddie Wood Endowment; the University of North Alabama Department of Geoscience Emma Craig Endowment; the University of North Alabama College of Arts, Sciences, and Engineering; and the University of North Alabama Center for Student Research, for funding assistance. We also thank Joe Weber and Selima Sultana for their valuable assistance and many suggestions for improving this chapter. Earlier versions of this chapter were presented at the Association of Pacific Coast Geographers conferences in 2017 and 2018, at the California Geographical Society conference in 2021, and at the American Association of Geographers conference in 2018, and we thank the audiences at those conferences for their comments.

Declarations

Conflict of interest The authors have no conflict of interest to declare that are relevant to the content of this chapter.

References

CNN. 2015. Does the National Park Service have a youth problem? *Cable News Network*, 19 March 2015. Accessed online 24 October 2019.

Duda, M.D., S.J. Bissell, and K.C. Young. 1998. *Wildlife and the American mind: Public opinion on and attitudes toward fish and wildlife management*. Harrisonburg, VA: Responsive Management.

Hungerford, H., and T. Volk. 1990. Changing learner behavior through environmental education. *Journal of Environmental Education* 21 (3): 8–21.

National Park Service. 2023. 2022 *Socioeconomic monitoring of National Park Service visitors*. Natural Resource Report NPS/NRSS/EQD/NRR—2023/2550. Washington DC: National Park Service. https://irma.nps.gov/DataStore/DownloadFile/691038

Pergams, O., and P. Zaradic. 2006. Is the love of nature in the US becoming the love of electronic media? 16-year downtrend in national park visits explained by watching movies, playing video games, Internet use, and oil prices. *Journal of Environmental Management* 80 (4): 387–393.

Pergams, O., and P. Zaradic. 2008. Evidence for a fundamental and pervasive shift away from nature-based recreation. *Proceedings of the National Academy of Sciences of the United States of America* 105 (7): 2295–2300.

Ruiz, T.S., and D. Havlick. 2023. A case study of Millennials' attitudes toward U.S. national parks. *Journal of Outdoor Recreation, Education, and Leadership* 15 (1(special issue)): 29–45.

Statistica.com. 2024. Population distribution in the United States by generation. https://www.statista.com/statistics/296974/us-population-share-by-generation/#:~:text=In%202023%2C%20Millennials%20were%20the%20largest%20generation,up%20about%2021.71%20percent%20of%20the%20population. April.

The Economist. 2013. Why go outside when you have an iPhone? *The Economist*, 17 August 2013. Accessed online 24 October 2019.

Twenge, J. 2023. *Generations: The real differences between Gen Z, Millennial, Gen X, Boomers, and Silents—and what they mean for America's future*. New York: Atria Books.

United States Census Bureau. 2023. U.S. population expected to begin declining in second half of century. Press Release CB23-189. https://www.census.gov/newsroom/press-releases/2023/population-projections.html.

United States Congressional Budget Office. 2024. The demographic outlook: 2024–2054. https://www.cbo.gov/publication/59899, January.

Washington State University, Social and Economic Sciences Research Center. 1982–2017. National Park Service projects. https://sesrc.wsu.edu/nps/. Accessed June 2024.

Weber, J., and S. Sultana. 2024. *The parks belong to the people: The geography of the national park system*. Athens: University of Georgia Press.

Xiao, X., K.J. Lee, and L.R. Larson. 2022. Who visits U.S. national parks (and who doesn't)? A national study of perceived constraints and vacation preferences across diverse populations. *Journal of Leisure Research* 53 (3): 404–425.

Index

J. Weber and S. Sultana (eds.), *The Changing Geography of National Parks and Protected Areas*,
https://doi.org/10.1007/978-3-031-74653-6